FPGA Design Beyond Logic
Managing Constraints

James E Moran

Principal Member of the Technical Staff
Draper Laboratory

Senior Adjunct Professor
University of Massachusetts Lowell

Table of Contents

Table of Figures

Index of Tables

Introduction

Engineers tend to focus on logic design when developing Field Programmable Gate Arrays (FPGAs). They write good code using a Hardware Description Language (HDL) that matches an agreed-upon set of requirements. Using a Hardware Verification Language (HVL), an independent verification team ensures that the HDL code matches these requirements. Throughout the process, a suitable source control system ensures the integrity of the design and verification code. The logic design is complete.

When these tasks are complete, there is a perception that the designer can press the button in the manufacturer's design software, wait a couple of hours, download the result to the target system, and everything works. If the physical characteristics of the FPGA and the surrounding system are not considered, the design rarely works on the first try. Now is a good time for a definition of what engineers do.

> Engineers apply mathematics, scientific principles, and state-of-the-art material to find novel solutions to problems or to improve existing solutions. An engineer's task is to identify, understand, and interpret the *constraints* on a design to yield an optimal result.

In FPGA design, constraints are contained in additional source files that inform the synthesis and place and route (implementation) tools on optimally creating a working FPGA using the selected device's physical properties and the physical properties of the surrounding system into account. It is important to note that the constraint file is a source file that is at least as necessary as any of the HDL files that define the logic of the system.

I gave somebody advice that I was surprised he (my son) followed. I told him that you do not need to think outside the box until you have used up everything inside the box. With FPGAs, you buy "the box" when you select the component. To be successful, you need to know how to make trade-offs between the Temporal Constraints and Spatial Constraints of "the box". The task of creating a successful FPGA design ends up being the optimization of multiple components of the system.

This text contains eight chapters. Chapter 0 describes the steps to develop good code to describe the FPGA logic. It is expected that those steps are completed before entering into the physical design phase. Chapter 1 and Chapter 2 describe Temporal Constraints and Spatial Constraints, respectively. These two chapters are the primary focus of the book. Chapter 4 describes steps used in the physical design process to achieve a successful FPGA design by optimizing the constraints and the original HDL if necessary, Chapter 5 describes good hardware design practice for FPGA Design. A description of how to stage the FPGA toolchain is shown in Chapter 6. To put things in context, Chapter 3 and Chapter 7 are case studies of how applying good design practice can lead to a successful FPGA design.

0.   Good Design Practice

A student or working professional with the desire to excel in the discipline of FPGA design should complete a course in FPGA design that either provides a strong background in Hardware Design Languate (HDL) or take a separate class focusing on HDL design. It does not end there. Even if there is no desire to work as a verification engineer, an FPGA designer should have formalized knowledge of the verification process and languages. The good news is that HDL and Hardware Verification Languages (HVL) frequently use the same language. The HDL subset, called synthesizable HDL, uses different constructs than the behavioral subset of the language used as an HVL. It does not end here, either.

To be disciplined in the engineering process, a course in software engineering is in order. Software engineering will describe language construction, good design processes, and coding styles. In addition, a software engineering course should include a component dedicated to source control tools and practice. Take the software engineering course from the Computer Engineering department, not the Computer Science department. Software engineering courses also include lectures on requirements engineering and test planning, which are valuable skills for any engineering discipline.

FPGA manufacturers also provide design software for their parts. To be successful, an engineer must be familiar with the latest version of the manufacturer's software that supports the selected component.

0.1 Design Specification

A well-written design specification made up of requirements that are not open to interpretation provides a single source of truth for all stakeholders. The more time spent getting the requirements correct, meaning matching the stakeholders' needs, the higher the chance of success. Developers can measure success in several ways.

> - Does the design match what the user needs? It is essential to understand who the user is. In FPGA design, the component user is not the system's end user. The user is the PCB designer or system integrator.
> - Does the design match what the manufacturer needs? An improperly constrained design may work under some ambient conditions like typical temperatures, typical power supply conditions, and average process results. Stretching these parameters can cause failures to occur during manufacturing, or worse, in the field. These failures will show up as yield issues and will make the product appear to be unreliable.
> - Finally, does the design match what your organization needs? Your organization needs to be profitable. In business school, they would define this as selling something for more than it costs. The cost of materials is easy to measure. The cost of design errors and manufacturing yield problems are harder to measure because they have a measurable effect on the bottom line but can also cause a loss of confidence in FPGA users and the target system's end users.

These items are a partial list of how success is measured. Early in the design process, developing a multifaceted view of what would be considered a success for a given FPGA design is helpful.

The mechanics of the logic design process are worth introducing but be aware that plenty of works (See Bibliography) address these topics in more rigorous detail.

0.2 Coding Style Practice

Coding guidelines are arbitrary. A complete set of arbitrary coding guidelines will build consistency into the code. Coding guidelines are essential when numerous designers work on the same design or multiple verification engineers work on a single verification test suite. These guidelines (and I'm careful to call the guidelines, not rules) define how you write a snippet of HDL code from which you expect the synthesizer to imply your logic. I suspect there are about five ways of writing code that imply a register, all of which produce a correct result. A coding guideline should be written that describes one version of the HDL code that implies a register to be used by every designer when they want to imply a register. This guideline will make the code consistent, easy to read, and easy to maintain. Figure 1 depicts examples of code that implies a D Flip-Flop in SystemVerilog and VHDL . A similar type of guideline should be written for every synthesizable construct that is available to the designers. Examples include multiplexers, counters, shift registers, and finite state machines.

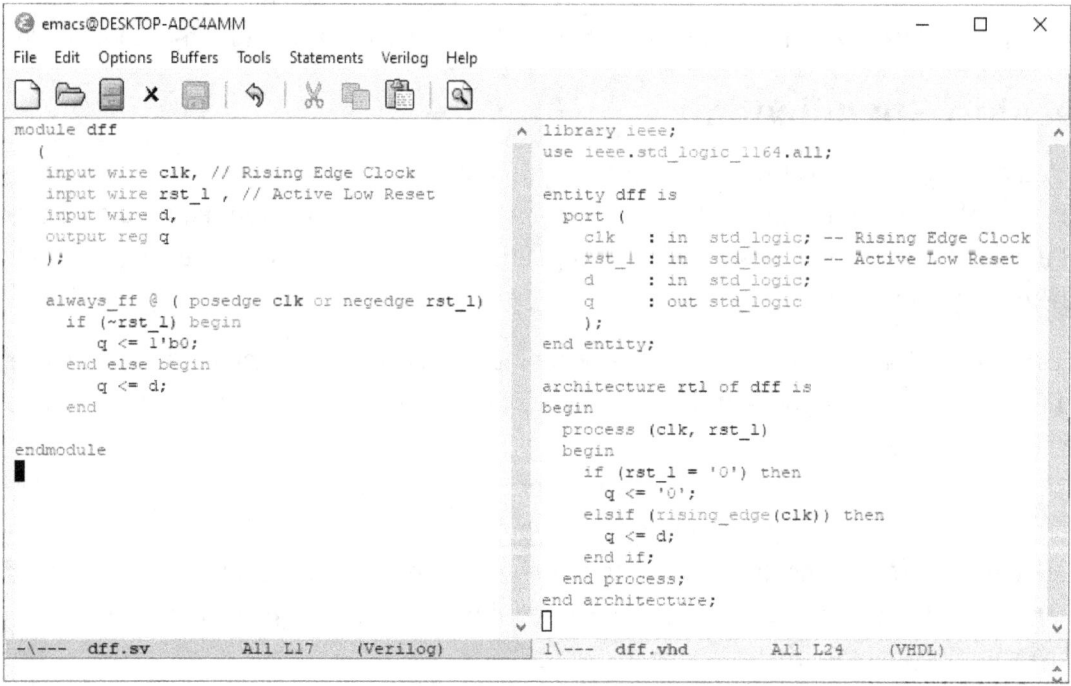

Figure 1: D Flip-Flop Example

Do not write the coding guidelines from scratch. Plenty of sources in textbooks, the internet, and legacy corporate material allow you to create your own coding guidelines. A common set of guidelines should be used in development for a given FPGA project or even as a corporate standard for every FPGA, and potentially Application Specific Integrated Circuit (ASIC) design. The description so far defines specific constructs, These guidelines should include minutia like character case, the number of spaces to indent, and how vertical white space is managed. A context sensitive editor can help enforce coding guidelines.

The twin sibling of coding guidelines is naming conventions. These are also guidelines, not rules, of how names are built. They apply to signals, modules, functions, filenames, etc. The naming conventions should include a format for how things are referred to. For example, if you have a serial port, the guideline should be to name signals with the prefix "sp_". This prevents some designers from referring to it as "serial-port_", "serial_port_" or not including a reference. More generally, the actual function of a signal should also have a designator. If a signal is a counter, it should have "_count_" (or "_counter_" or "_cnt_"). The actual name is not essential, but consistent usage is necessary. Input and output signals should have an "_in" and "_out" suffix (or spell it out but be consistent). Active low is "_l" or "_n". The suffix "_l" is preferred because differential signals are frequently suffixed with "_p" and "_n". Two important named signals in a design are the clocks and resets.

Every clock signal in the design should include "_clk_" and the string "_clk_" is never used on a signal that is not used as a clock, this designation will make it easy to extract the clock signals from your design. This will be important for the description of clock constraints in Section 1.1. Similarly essential is the reset signal. A designation like "_rst_", if used like the "_clk_" string above, allows the reset signals to be uniquely identified. Although there is no specific constraint for this, it is a helpful practice to track which reset signals are associated with a given clock. For example, a reset signal in the clock domain of clk_1 should be rst_l_1.

In general, filenames should be named after the module or entity that is contained in them. For example, a SystemVerilog file containing a module called dff should be called dff.sv. A file should only contain a single module or entity.

Experienced designers understand the benefit of coding guidelines and how to use them effectively. When assembling a team from diverse backgrounds, the designers may use different guidelines (because they are arbitrary). A program manager's challenge is merging these guidelines into a single set that the team can use consistently. In addition, I mentioned that these are guidelines, not rules implying that there could be exceptions. Exceptions should be precisely that. Exceptions are not be granted just because a designer does not want to follow the guidelines. Exceptions require formal waivers and are be granted for cases where following the guideline is impossible or would cause an inconsistency.

I have seen coding guidelines that were hundreds of pages long. These documents do not contribute to the design process because they are not a reference a designer can use during the code construction process, and they surely cannot memorize all the content. I have maintained and supervised the use of a thirty-page coding guideline document. That was more tenable because it is something that a disciplined designer can refer to during code construction and can memorize the most critical parts. I believe a coding guidelines specification should be on two sides of a letter-sized page, laminated for good measure. A single-page specification gives the designer a quick reference to a small set of guidelines, and compliance with the guidelines will be higher than the longer documents.

0.3   Coding for Synthesis

Section 0.2 suggested that each digital logic construct would have a corresponding code snippet to imply it. What if there is no logic construct available to perform that function? The implication of logic constructs is where FPGA design practice and ASIC design practice differ. For example, FPGAs do not contain latches, ASICs do. Latches are synthesized from available logic, but it is not good practice. FPGAs do not have internal three-state buffers, ASICs do. Make a note in the coding guidelines warning the user not to use internal three-state buffers. These limitations generally do not exist in mainstream logic design but, special constructs like latches and three-state buffers should be noted in the coding guidelines.

In addition to what is or is not available in the FPGA, coding for synthesis should follow certain design practices based on the FPGA architecture. FPGA logic elements are built with combinational inputs and registered outputs. For this reason, design constructs in your code should be written with combinational inputs and registered outputs. An exception to this is on the inputs to the chip itself which should be registered. IO cells in an FPGA have local registers. For this reason, the inputs to the FPGA should not be combinational and should be registered. A schematic showing this function is shown in Figure 2.

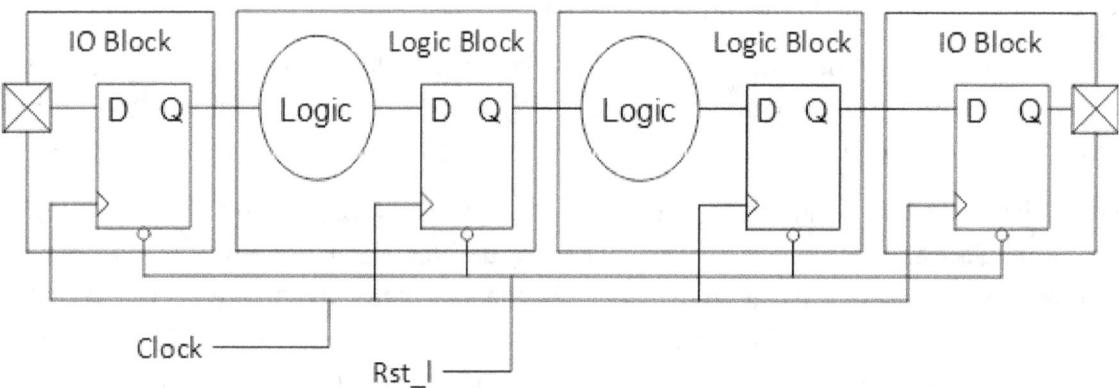

Figure 2: Architecture for FPGA Synthesis

While we are discussing the concept of constructing logic to fit an FPGA architecture, the actual architecture of a finite state machine (FSM) should be visited. You probably heard of Mealy and Moore state machines at some point. They are valid concepts, and it is always good to standardize on a logic format. The basic idea of Mealy and Moore deals with whether inputs and outputs are registered or not. Given the architecture shown in Figure 2, it is implied that a logic design to be targeted to an FPGA should have combinational inputs and registered outputs. This is similar to a Moore state machine but the output does not need to be the value of a state. If the inputs to a finite state machine come from the inputs of the device, an additional Flip-Flop stage should be inserted between the pins and the state machine. A template for an FSM should be included in the coding guidelines.

14

In addition to following the FPGA architecture to improve synthesizeability, clock domain crossing and reset synchronization are handled using industry-accepted methods. The specifics of these concepts are detailed in Chapter 1

0.4 Design Verification

Verification is a process that shows that a design matches the requirements listed in the design specification. The verification process does this by creating a suite of tests. Each test verifies that the operation of the code matches what is stated in a given requirement or requirements in the design specification. Once the requirements are verified, the design is thoroughly verified. The verification flow is outlined in Figure 3. Because the number of states in an FPGA is large, the verification task frequently uses techniques where the stimulus is random data constrained around a specific range. The response of the device is measured against an expected response. To ensure that the device is being verified effectively, a measure of coverage is made during the test case. Develop directed test cases to cover areas difficult to cover by constrained random stimulus tests.

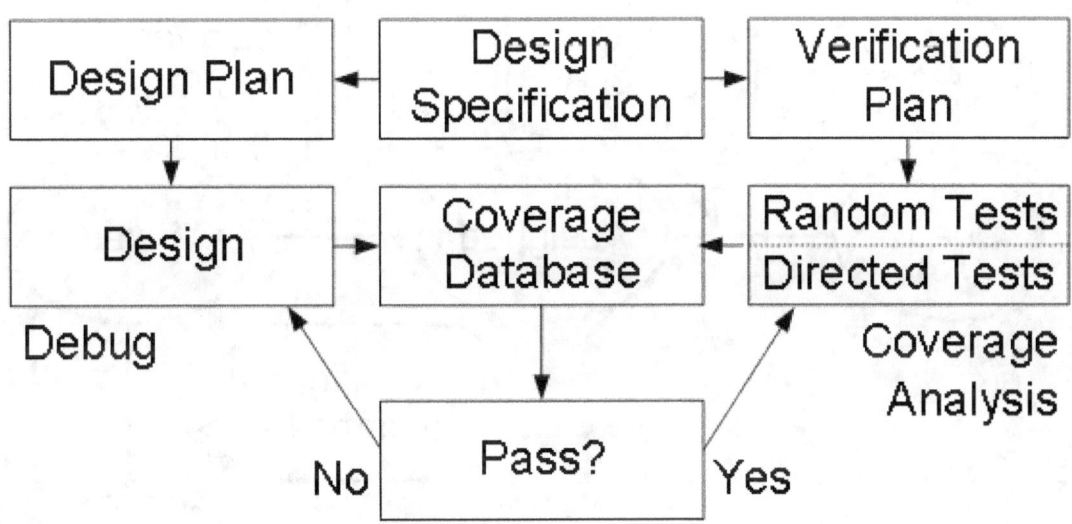

Figure 3: Verification Flow

0.5 Configuration Management

Configuration Management allows design engineers and verification engineers to track the creation, modification, and release of source files of the design. The obvious choice for these files includes the synthesizable code and the verification testbench. A vital configuration item pertinent to this discussion is the constraint file. This "minimum set" of configuration items is used to reconstruct the design. The design specification is also essential source information and is usually tracked in Configuration Management. Machine-generated files like the compiled description of the chip or simulation results are generally not configuration-managed because of their size and the fact that they are regenerated from source information. Specific versions of design tools should also be stored as source information because it may be necessary to recreate an FPGA image at a later time with the exact version of the design tool that it was initially created from.

An additional feature of Configuration Management is bug tracking. This feature allows, for example, a verification engineer to write a test that finds an anomaly (bug) and reports it. This report puts the bug into an open state. The bug is assigned to the appropriate designer who disposes of the bug by either changing the code to fix it, changing the design specification because it did not describe the correct behavior, or explaining to the verification engineer that the bug is not an anomaly. The designer sets the bug to a fixed state. The bug report is then returned to the verification engineer, who retests the system and decides if the bug has gone away. The verification engineer will also run all previous test cases (regression testing) to be sure that the fix to this bug did not cause a different bug to appear in an area previously tested and shown to meet the design specification. Once this is done, the bug is set to a closed state. A simplified lifecycle of a bug is shown in Figure 4.

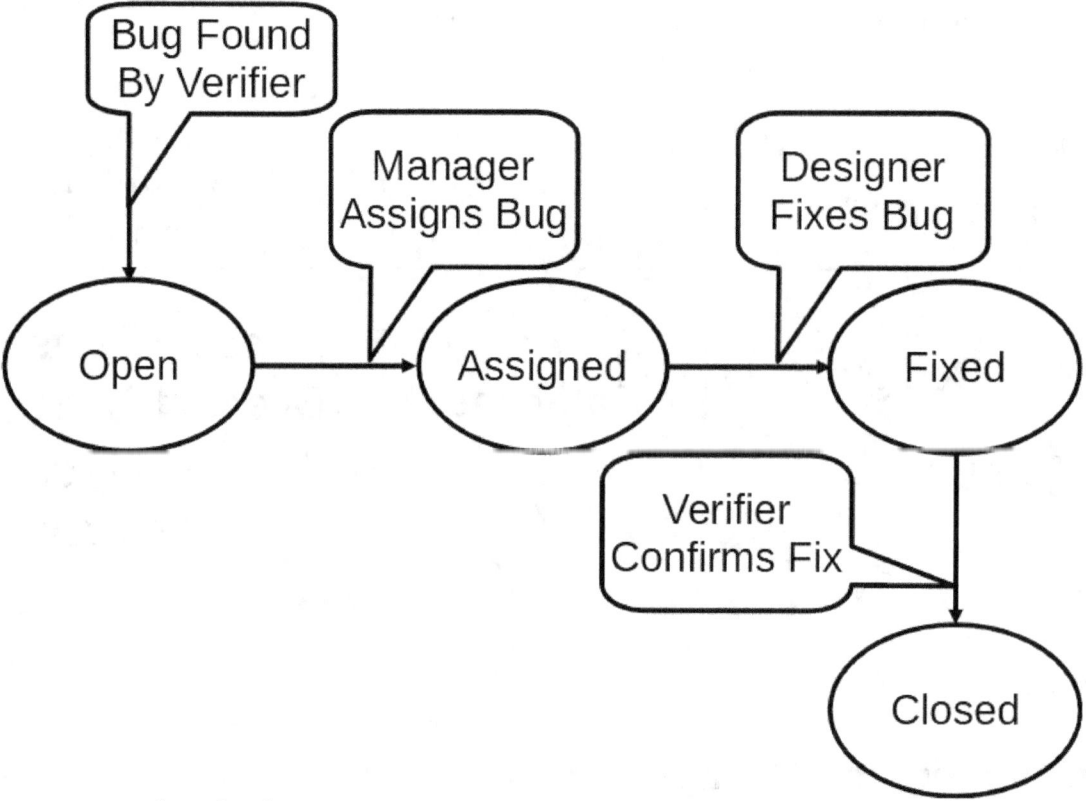

Figure 4: Lifecycle of a Bug

1. Temporal Constraints

Temporal Constraints define the limits for describing the timing requirements for the design.

In general, constraints relay physical information to the synthesizer to bias the final result. Always treat constraints as source information. A design implemented with missing constraints may work but can show unexplained failures under certain conditions and also perform differently if small changes are made. An over-constrained design may fail to achieve timing closure but have the potential to be a functional design. It is essential to constrain everything that needs to be constrained but little else.

1.1 Clock Constraints

To fully describe clock constraints, a few terms must be defined. These are generally referred to as the AC Characteristics.

- Frequency (F) is the desired minimum frequency system. It must be less than F_{MAX}.
- Period (T) is the clock period. It is equal to the reciprocal of the Frequency.
- Setup Time (T_{SU}) is the minimum time data to a Flip-Flip must be stable before the active edge of the clock to the Flip-Flop.
- Hold Time (T_H) is the minimum of time data to a Flip-Flip must be stable after the active edge of the clock.
- Clock to Q (T_{CK2Q}) is the maximum time after the active edge of the clock before the Q output of a Flip-Flop is stable

Clock constraints inform the synthesis and place and route tool of the maximum frequency of a given clock domain in the FPGA. The static timing analyzer calculates each path's maximum frequency (F_{MAX}) in a given clock domain. The path with the slowest F_{MAX} must be faster than that domain's frequency (F) clock constraint. F_{MAX} for a path is calculated using clock input to Q output (T_{CK2Q}) of the sending Flip-Flop, propagation delay (T_P) of the logic between the sending Flip-Flop and the receiving Flip-Flop including any path delay, and the Setup Time (T_{SU}) and Hold Time (T_H) of the receiving Flip-Flop. F_{MAX} is calculated using Equation 1, below. It should be noted that the F_{MAX} equation shown below assumes there is no clock skew meaning the clock is received by both the sending and receiving Flip-Flop simultaneously.

$$F_{MAX} = \frac{1}{\left(T_{CK2Q} + T_P + T_{SU}\right)} \qquad (1)$$

The actual period of the clock is $T_{CK2Q} + T_P + T_{SU}$. The minimum period (T) minus the actual period is defined as slack. Figure 5 shows a model used for timing analysis in cases with is no clock skew.

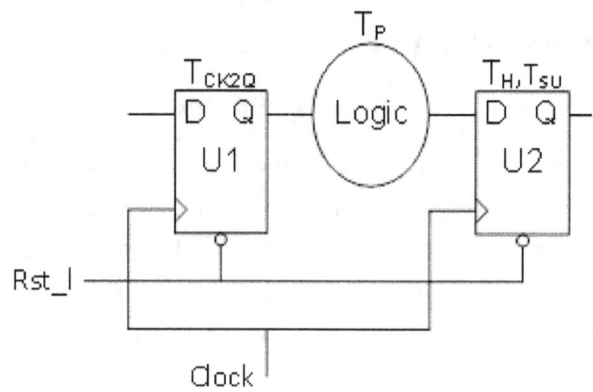

Figure 5: Model for Timing Analysis

A timing diagram associated with the model is shown in Figure 6.

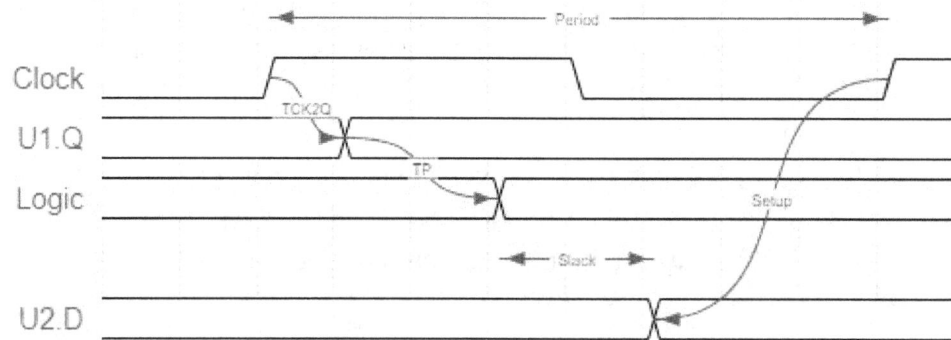

Figure 6: Waveform for Timing Analysis

Slack Calculation Example:

For a given system, the desired clock frequency (F), which is the clock constraint set for the "Clock" signal in Figure 5, is 500 MHz. Given the following parameters, will the model shown in Figure 5 meet timing?

T_{CK2P} = 750 pS; T_P = 850 pS; T_{SU} = 300 pS; T_H = 0 pS.

$$F_{MAX} = \frac{1}{(750\ pS + 850\ pS + 300\ pS)} = 526\ MHz$$

This calculation shows that the maximum frequency of the model is 526 MHz. Because the target frequency is only 500 MHz, the model will achieve timing closure. Another way of looking at that is to calculate timing closure based on time. In the example above, the desired period is the reciprocal of 500 MHz which is 2000 pS. The period of the model is 1900 pS. In this case, there is 100 pS of margin, also called slack. If the slack is positive, the circuit achieves timing closure. If there is negative slack, there is a timing violation. Other phenomena can take time away from the margin. Signal noise and clock skew are common properties that reduce the margin. Also, the parameters are sensitive to temperature, process variation, and voltage, but the desired frequency is not. For this reason, the timing closure equation must be solved over temperature, process variation, and voltage, and the worst case is used to define timing closure. Note that different processes have different characteristics over temperature and voltage, so follow the manufacturer's recommendations.

A skew in the clock path, will cause the F_{MAX} of a given path to go up or down.

The F_{MAX} in equation 2, accounting for clock skew, is shown below.

$$F_{MAX} = \frac{1}{\left(T_{CK2Q} + T_P + T_{SU} - T_{SKEW}\right)} \qquad (2)$$

A negative skew, shown in Figure 7, is a case where the clock to the sending Flip-Flop is delayed.

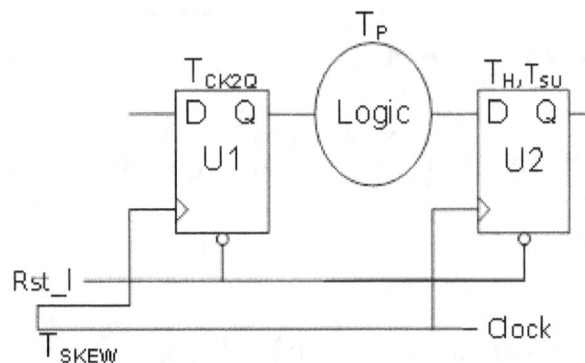

Figure 7: Negative Skew Timing Model

The timing diagram in Figure 8 shows that there is less slack than the case with no skew, which would cause F_{MAX} to go down. In practice, the worst-case negative slack (WNS) is used to calculate F_{MAX} for the clock domain is metric used during timing closure. Total negative slack (TNS) is commonly used to gauge the progress of the timing closure process. It is the sum of the negative slack values, usually reported with the total number of negative slack paths.

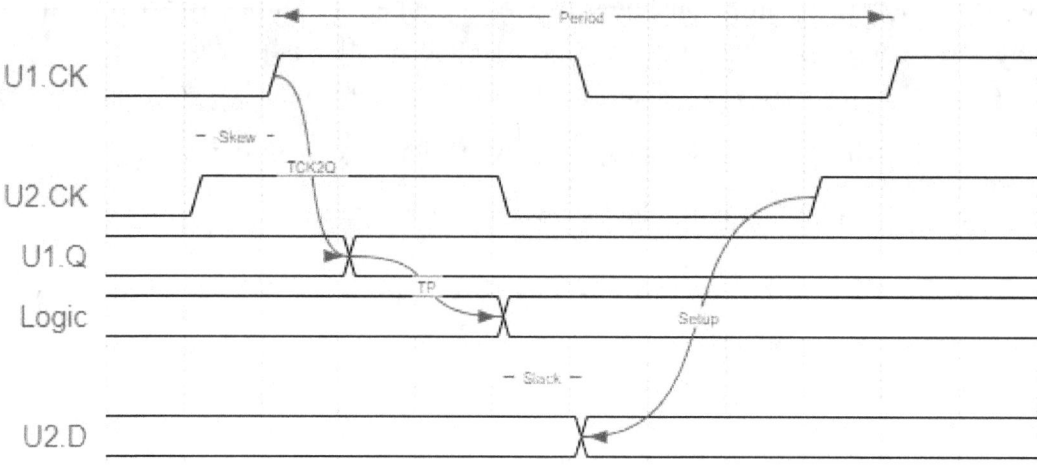

Figure 8: Negative Skew Timing Diagram

Negative Skew Example:

Using the same parameters as the Slack Example, the frequency is calculated assuming a negative skew of 185 pS. The calculation yields a frequency of 480 MHz. This frequency is a lot slower than the zero skew case and does not achieve timing closure.

$$F_{MAX} = \frac{1}{(750\ pS + 850\ pS + 300\ pS + 185\ pS)} = 480\ MHz$$

There are several ways to remedy a negative slack timing violation if F_{MAX} is below the target frequency. Reducing the skew is an obvious way to reduce the slack. Most FPGAs have clock management and routing resources to reduce skew (see section 1.7). If that does not remove timing violations, reducing the amount of logic between the Flip Flops will reduce T_P. Techniques for doing this are mostly centered around some type of pipelining (see section 1.9). If possible, using a part with a faster speed grade can reduce negative slack. The consequence of using a faster speed grade part is cost. Finally, and least desirable is to reduce the system's target frequency. Reducing the system's frequency will likely reduce the system's performance and may cause the system not to meet its requirements. The advantage of reducing the frequency, however, is lower power consumption. Section 1.9 contains more detail about timing closure.

Skew in the opposite direction can increase F_{MAX}. A positive skew, shown in Figure 9, is a case where the clock to the receiving flip flop is delayed by 185 pS.

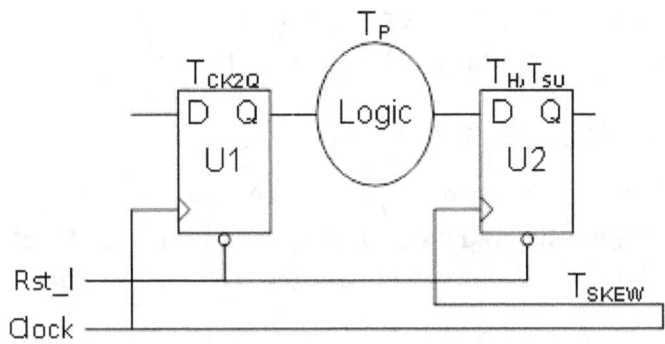

Figure 9: Positive Skew Timing Model

A timing diagram showing how positive skew can increase slack is shown in Figure 10.

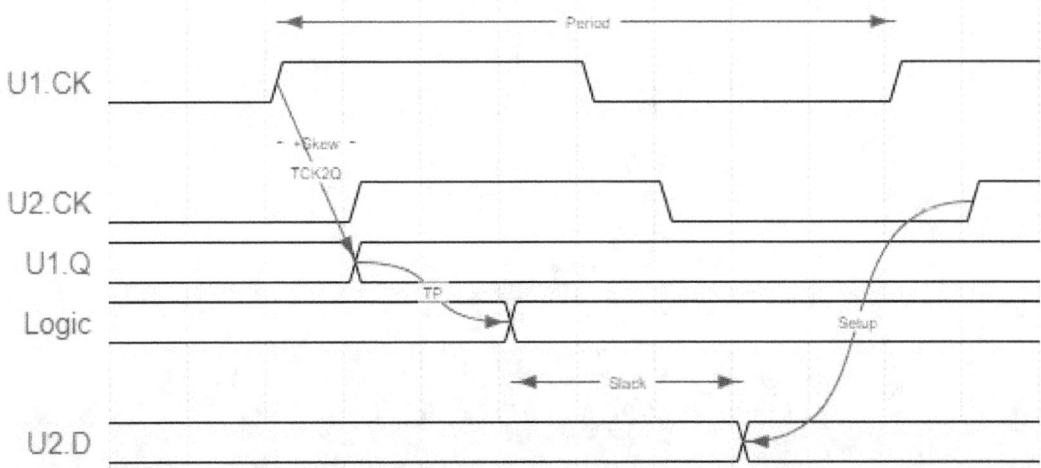

Figure 10: Positive Skew Timing Diagram

Positive Skew Example:

To repeat the example once again, the calculation yields 583 MHz. This frequency is faster than 500MHz and achieves timing closure. This margin could also be stated as 285 pS of positive slack.

$$F_{MAX} = \frac{1}{(750\ pS + 850\ pS + 300\ pS - 185\ pS)} = 583\ MHz$$

At this point, the reader may be thinking, "Why not add positive skew everywhere to increase the clock speed?" There are two reasons why that is not valid.

First, delaying a given Flip Flop with respect to the clock reduces the available period for the next stage. Given that digital systems are generally chains of logic separated by Flip Flips, it would slow down the F_{MAX} in the next stage, which would slow down the whole chain. The other reason is that too much positive skew can cause the signal to go away before the receiving Flip-Flop clocks it in. This is a hold time violation. In general, $T_{skew} < (T_{CK2Q} + T_P + T_H)$. This condition should be part of the timing closure calculation.

Generally, systems with many of paths have a combination of positive and negative skew. The minimum slack would define F_{MAX} for the clock domain in this case.

1.2 Input Output Timing Constraints

The setup time and hold time parameters of a given FPGA inside the fabric, as described in Section 1.1 are different than the setup time and hold time parameters of the device input-output (IO) pins. In addition, the static timing analysis is more of a manual process for the IO than inside the fabric.

Figure 11 shows a model where two FPGAs are connected on a PCB. There is a common clock and reset. The sending register, on the left, has a timing parameter (T_{CK2Q}) which is the maximum time from the clock input transition to the output changing on the FPGA. The T_P parameter is the wire delay on the board. It is calculated based on the trace length and dielectric constant of the board material using Formula 3 below.

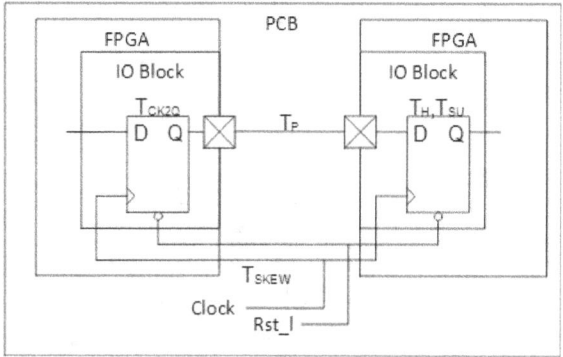

Figure 11: FPGA to FPGA Model

$$T_P = \frac{11.8}{\sqrt{\varepsilon_r}} \, inch/nS \tag{3}$$

In this equation, 11.8 is the speed of light in pS/inch, ε_r the dielectric constant of the board (usually between 3.8 and 4.8 for FR-4). In this case, the signal travels at about 5.4 in/nS which is 185 pS/in. This delay corresponds to the clock skew used in the examples in Section 1.1.

Given that there are likely many traces on the board, their lengths must be extracted from the PCB design file and the T_P for each trace must be calculated using Equation 3. An example of this is shown in Figure 12. In addition, the actual value for T_{CK2Q} must be extracted from the timing report for the sending FPGA and the actual value of T_{SU} must be extracted from the timing report for receiving FPGA. F_{MAX} for each of these paths using the shown in Equation 2. This calculation is likely done in a spreadsheet where each result is compared to the desired operating frequency (F) If the F_{MAX} of all of the traces is above the desired F_{MAX}, timing closure has been achieved. Of course, the clock skew must be taken into account to ensure there are no hold time violations. An example of notional delays, shown in Table 1, shows a group of signals connected between two FPGAs. The slack is calculated assuming a frequency of 200 MHz (5.0 nS period). The initial calculation is done assuming no skew. The Slack column shows positive slack between 1.0 nS and 2.3 nS. D2 is noted as the Worst Case

(WC). This is used as the maximum negative skew. Stated another way, if the negative skew is equal to or greater than 1.0 nS, there will be negative slack which equates to a timing violation. It would not be the best practice to assume a system had no slack, so it is practical to set the maximum skew to 0.7 nS – 0.8 nS (~4 inches). Also, these numbers must be evaluated over the full operating temperature, voltage, and process range.

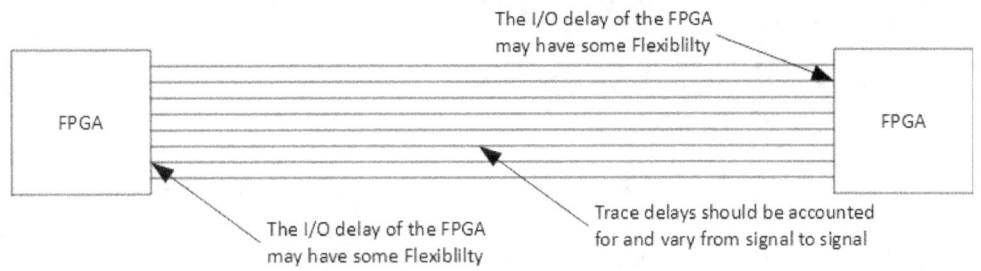

Figure 12: Bus Timing Example

Table 1: Bus Timing Values

Signal	T_{CK2O}	T_P	T_{SU}	Slack
D0	1.2 nS	1.0 nS	0.5 nS	2.3 nS
D1	1.2 nS	1.0 nS	0.5 nS	2.3 nS
D2	1.7 nS	1.8 nS	0.5 nS	1.0 nS (WC)
D3	1.5 nS	1.1 nS	0.5 nS	1.9 nS
D4	1.4 nS	1.6 nS	0.5 nS	1.5 nS
D5	1.3 nS	1.4 nS	0.5 nS	1.8 nS
D6	1.3 nS	1.0 nS	0.5 nS	2.2 nS
D7	1.8 nS	1.1 nS	0.5 nS	1.6 nS

Each output on an FPGA must have a constraint set for T_{CK2Q}. This constraint defines the maximum allowed time from the active edge of the input clock to the transition on the output pin with all skews taken into account. Each input on an FPGA must also have a T_{SU} and a T_H constraint set with respect to the input clock. Bidirectional connections must have both input and output constraints set. Notice that in the previous paragraph that I stated that the actual delays should be used. If the FPGA achieves timing closure, these will be less than or equal to the constraints, so the board-level timing analysis has a better chance of a positive result. It is important to note that even though on a given implementation run, the FPGA achieves timing closure the actual values can change. For this reason, the board-level timing analysis must be recalculated anytime either FPGA changes.

The above examples describe a system where an FPGA is connected to an FPGA. In this case, the output constraints of the sending FPGA and the input constraints of the receiving FPGA are adjusted to improve timing. In a case where there is an FPGA connected to an off-the-shelf component, such as a processor, memory, or ASIC, the timing characteristics for off-the-shelf components are fixed, and the FPGA must be adjusted to match them. An example of this is shown in Figure 13.

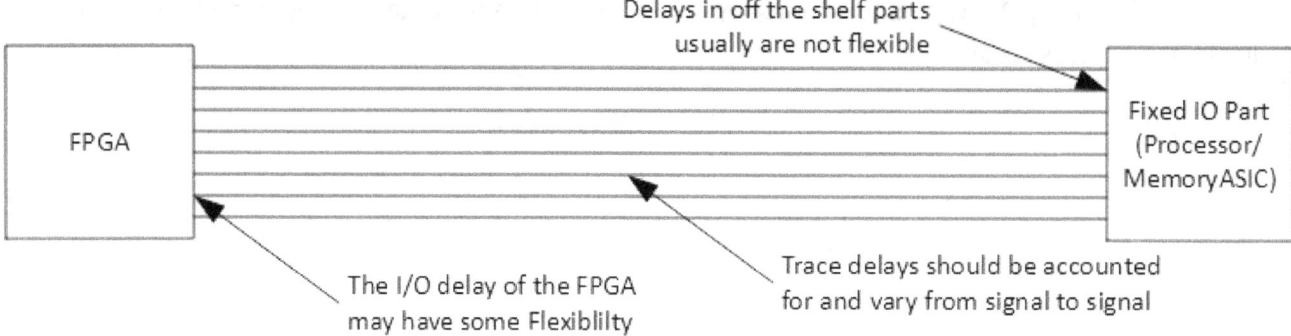

Figure 13: Timing closure to Fixed Timing Components

1.3 Over-constraining

There is a tendency for system designers to be very conservative. For example, setting a clock constraint for a system to 600 MHz when the requirement is only 500 MHz. Setting a constraint well beyond what it needs to be is called over-constraining. The perception is that the extra margin will improve reliability. The skew examples in Section 1.1 would now be timing violations. The designers of FPGA components, meaning the silicon manufacturers, have the proper margin built into their static timing analysis to take care of any margin.

For this reason, there is no need to set a constraint significantly above the desired value. Another thought is that some over-constraining could positively affect place and route. This is not the case and can slow down the run time of synthesis and place and route or cause a design that would otherwise have achieved timing closure to have timing violations.

1.4 Multicycle and False Paths

A multicycle path is a timing path where the circuit is designed such that although the signal is changing, the logic after the receiving Flip-Flip will not do anything with the signal on the next clock cycle. In this case, if there is a timing violation, it will not affect F_{MAX}. Usually, a designer will look through the timing violations to check to see if a timing violation is on a multicycle path. If it is, there is a constraint to designate that path to be a multicycle path, and it is removed from the timing report. The constraint can also allow the designer to specify how many extra clock periods are necessary. Sometimes designers will make quick judgments on multicycle paths that make the design achieve timing closure when it does not. To be a multicycle path, a timing path must be logically verified not to require the new value of the signal on the next cycle and that the actual time delay is less than the number of cycles when it will need the new value of the signal. Multicycle path constraints need to be reevaluated anytime the logic changes.

Figure 14 shows an example of a multicycle path. The multiplexers feeding each Flip-Flop regulate when data is clocked into the Flip-Flop. Each Flip-Flop can change once every four cycles. The second stage is two cycles out of phase from the sending one. If this did not make timing on a single cycle, a constraint could be set to make it a multicycle path.

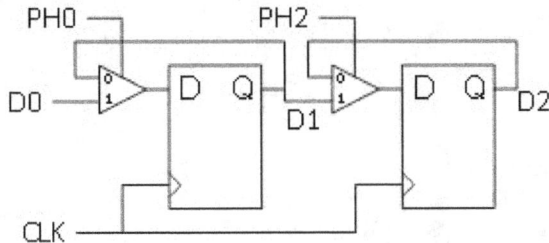

Figure 14: Multicycle Path Example

The timing diagram demonstrating this is shown in Figure 13. The four clock phases, PH0-PH3, PH0, and, PH2 are controlling the multiplexers feeding the Flip-Flops.

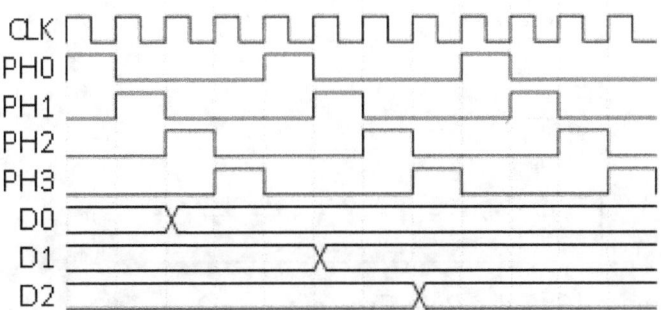

Figure 15: Multicycle Waveform

A false path is a case where a signal will never cause the input of the receiving Flip-Flop to change. The examples, so far, show a single Flip-Flop sending to a single receiving Flip-Flop. In cases where a group of sending Flip-Flops are connected to a group of receiving Flip-Flops through a combinational logic circuit, it is not out of the question that a given sending does not affect a given receiving Flip-Flop. This is an example of a false path and there is a constraint that can remove that from the timing analysis. As with a multicycle path, False paths should be reevaluated if the design changes.

Figure 16 is an example of a false path. Looking at the function, S_3 is a function of all of the inputs. S_0, however, is only a function of A_0 and B_0. In this case, The paths from S_0 and A_3, A_2, A_1, B_3, B_2, and B_1 could be constrained as false paths. This type of function is typical of an adder.

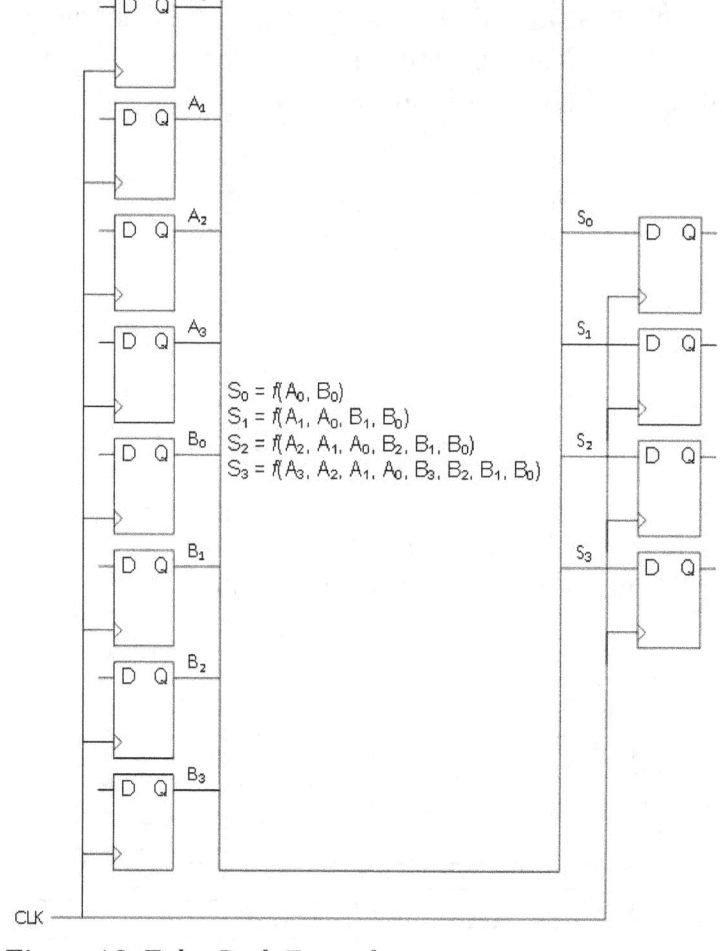

Figure 16: False Path Example

1.5 Multiple Clock Domains

It is frequently the case that a design contains multiple clocks. The sequential logic associated with a given clock signal is called its clock domain. When there is more than one clock, a signal will likely go from one clock domain to another. When this happens, special handling is required. Figure 17 is an example of a Clock Domain Crossing (CDC). The reset signal is shown for completeness.

The proper way to handle reset in a multi-clock system is described in Section 1.6.

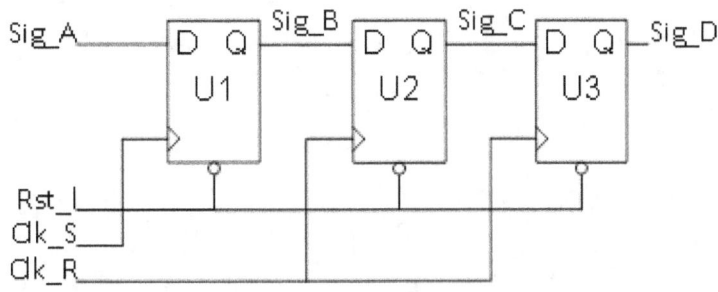

Figure 17: Clock Domain Crossing Synchronization

A timing diagram of a clock domain crossing is shown in Figure 18. If Sig_A is the input signal in the Clk_S (Sender) domain, Sig_A meets the T_{SU} and T_H requirements for U1. The result is Sig_B which occurs after T_{CK2Q} after the active clock edge, as shown in Figure 18. If the change on Sig_B happens coincidentally with the active edge of Clk_R, a metastable event can occur. This event is a small spike on Sig_C shown in Figure 18. This event can persist as an oscillation, causing additional improper operation throughout the design. To prevent this oscillation, a second Flip-Flip (U3) is inserted and connected to Clk_R. Although the metastable event can propagate through U3, it is improbable. In high-reliability systems, a second synchronizing Flip-Flip is added to reduce the probability of the metastable event propagating to the rest of the design. Note that the Sig_D signal may go high on the first edge or second edge of the Clk_R signal. This is expected behavior.

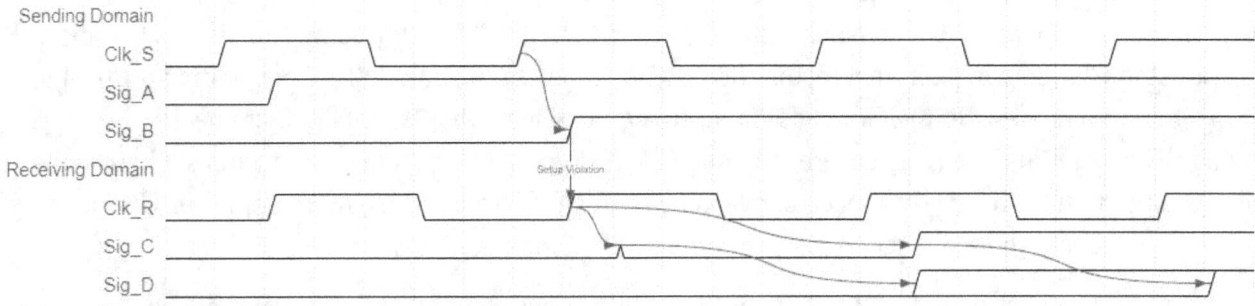

Figure 18: Clock Domain Crossing Timing Diagram

1.6 Reset Synchronization

The reset signal is the other important signal in the design. It must be properly synchronized to ensure that all of the Flip-Flops in the design see the reset signal during the same clock period. An example of reset synchronization is shown in Figure 19. The circuit within the dashed block shows a circuit that synchronizes an active low reset to a single rising edge triggered clock domain. Note that the clock is the same on all Flip-Flops. A unique feature of this circuit is that the D input of U1 is connected to a logical 1. When the Rst_l is brought to a logical 0, U1 and U2 go to a logical 0, causing the system to reset, regardless of the clock. When Rst_l is brought to a 1, the logical 1 is propagated through U1 and U2. If Rst_l is changed too close to the active edge of Clk_1, the Q output of U1 may temporarily enter a metastable condition. U2 will block any metastable condition. U3 is an example of a Flip-Flip in the Clk_1 domain. There are likely hundreds or thousands of Flip-Flops in a clock domain. U1 and

U2, wired as shown, will ensure that all Flip-Flops in the clock domain see the Rst_l transition in the same Clk_1 period as long as the propagation time of the reset signal is less than one clock period.

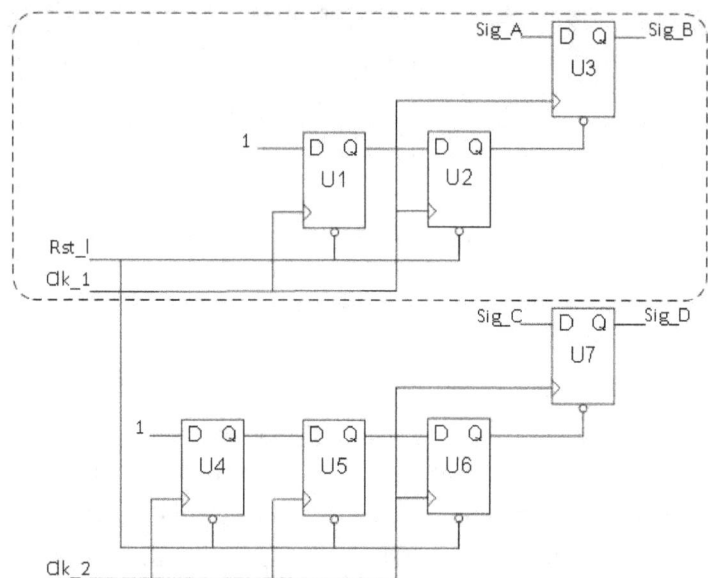

Figure 19: Reset Synchronization

Although there can be more than one clock domain, there should only be one reset signal. With that said, the single reset must be synchronized to each separate clock domain. The circuit in Figure 19 is made up of U4, U5, and U6 and synchronizes Rst_l to Clk_2. This is the same function as U1 and U2 except using Clk_2. An additional feature of the Clk_2 synchronizer is U6. This additional Flip-Flop is used to delay the reset to the Clk_2 domain. In a case where Clk_1 and Clk_2 are nearly the same, having the extra Flip-Flop (U6) ensures that the Clk_1 domain always emerges from reset before the Clk_2 domain. If there is a difference between the two clock frequencies, more copies of U6 may be necessary to ensure the sequence of reset throughout the clock domains.

Extra care in the construction of the reset circuit ensures consistent behavior every time the circuit is powered on and improves overall reliability.

1.7 Clock Management

Clock management is used to minimize clock skew within a clock domain. FPGAs contain specific components like Phase Locked Loops (PLLs), Clock Drivers, and Clock Distribution Networks that facilitate minimal skew clock distribution. Stated another way, clock management's function ensures that every Flip-Flop in a clock domain sees the active edge of the clock at as close to the same time as possible. Clock managers can configure the output clock's frequency, phase delay, and duty cycle. Frequency, phase delay, and duty cycle are illustrated in Figure 20. Clock Frequency is the reciprocal of the clock period, which is the time between active edges of the clock signal, as shown in Clock_1. Only one edge (positive) of the clock should be considered active. The phase delay is the time from an active edge of one clock with respect to another, usually measured in degrees. Figure 20 shows a 90°

phase delay between Clock_1 and Clock_2. Finally, the duty cycle is the percentage of time a clock signal is high relative to its period measured in percentage. Clock_3 in Figure 20 illustrates a 75% duty cycle.

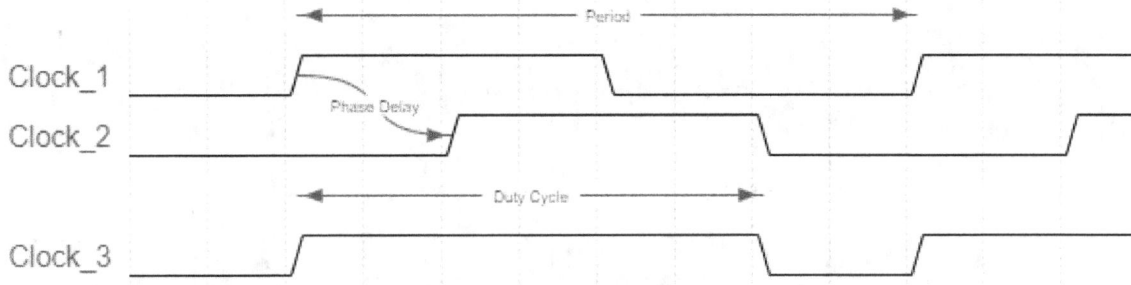

Figure 20: Clock Frequency, Phase Delay and Duty Cycle

Also, clock managers can safely switch between input clocks without violating clock parameters. A generalized block diagram of clock management in an FPGA is shown in Figure 21. A PLL is a circuit that varies the speed of its output clock to minimize the phase difference between the input clock and the clock fed back from the clock distribution network. Many works describe the theory and design of phase-locked loops. The actual design of the PLL is proprietary to the FPGA component manufacturer. The large triangle after the PLL is a clock driver. This is a high-capacity driver that can maintain the integrity of the clock signal across the clock distribution network. Clock drivers are high-power devices on the FPGA, so they should be used sparingly.

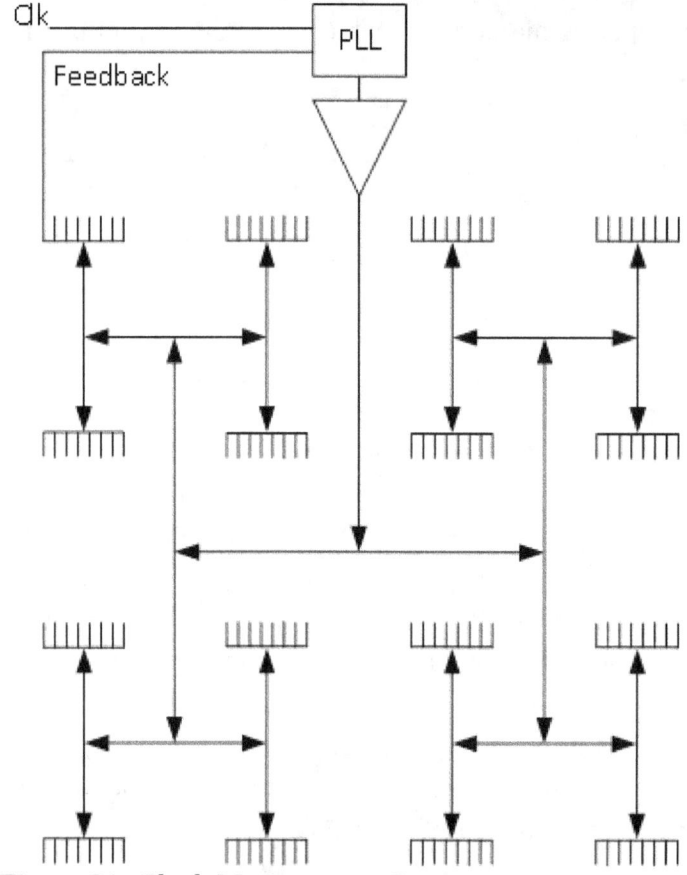

Figure 21: Clock Management Components

Finally, the clock distribution network in this example is an H-Tree. Looking at this network, you can see it is a fractal structure of repeating H's. Following any path from the driver to a destination takes the same distance, thus reducing the clock skew. The combination of these components will ensure the clock going to each destination switches with the minimum possible skew while maintaining the highest possible signal integrity.

1.8 Derived Clocks

Derived clocks are dangerous and should not be used. There are two classes of derived clocks. Clocks that are frequency divided using gates and Flip-Flops, and clocks that are switched using logic gates. The clock management block should be used in both cases to create the proper clock signal. It is worth showing an example of each type of derived clock to clarify what should not be done.

A ripple counter, as shown in Figure 22, is an example of a circuit that derives a clock using a Flip-Flop instead of a clock manager. The main problem with a ripple counter is the delay induced by the T_{CK2Q} of the Flip-Flop. This causes a phase delay between the input clock and the derived clock. Because T_{CK2Q} is variable over temperature, voltage, and process variation, the phase delay will also vary. The phase delay increases if there are multiple stages in the ripple counter,. The phase delay is exacerbated in an FPGA because of the flexible routing architecture. FPGAs generally have plenty of clock manager components, so using ripple counters or any other logic arrangement to derive different frequency clocks should not be necessary.

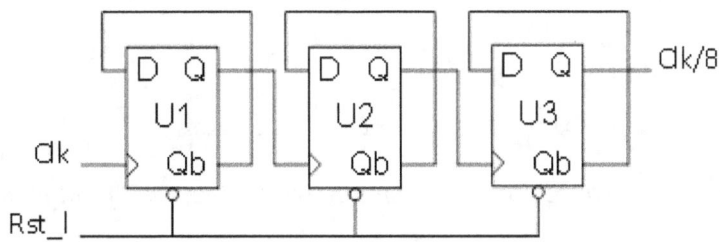

Figure 22: Ripple Counter Clock Divider

A timing diagram for a Ripple Counter Clock Divider is shown in Figure 23. The transition edges of Clk/8 are starting to approach the active edge of the Clk signal. If a signal from the Clk domain is clocked into the Clk/8 domain, it will need to be treated as a clock domain crossing. Also, if the T_{CK2Q} of the Flip-Flops starts to increase or there are more ripple stages, the edge could end up being after the next active edge Clk. Both of these conditions are very sensitive to temperature, voltage, and process variation, which could cause erratic behavior or complete failure. If the Clk frequency is lower, it might seem that the skew is less of a problem. That may be the case, but the variation can still cause a problem.

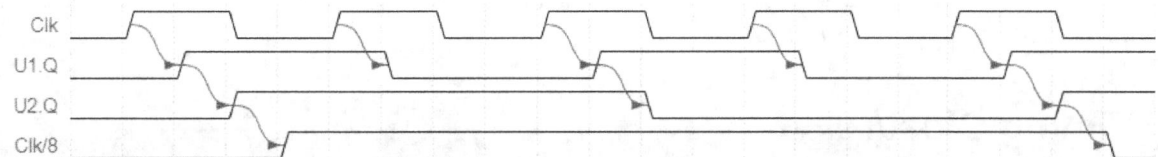

Figure 23: Ripple Counter Clock Divider Timing Diagram

It may be tempting for a designer that needs to switch between two clocks to use a simple 2:1 multiplexer. This can wreak havoc on the derived clock during the switching. If the switch happens too close to a clock edge, the output clock can have a runt pulse. This small pulse can cause T_{SU} violations and cause the circuit to be unpredictable. A literature search will show various implementations of a so-called glitch-less multiplexer. These work well in an ASIC where the delays of the cells and the lengths of the routes between them can be balanced. In an FPGA, the synthesizer and place and route tools may not place the components in the flexible architecture to be controlled in ways that balance their delays, and the glitch-less multiplexer may not be reliable. FPGAs generally have many clock manager components that contain clock enables and switches, so it should not be necessary to use glitch-less multiplexers or any other logic arrangement to switch between different frequency clocks. An example showing a runt pulse caused by a clock switching from on to off using common components is shown in Figure 24.

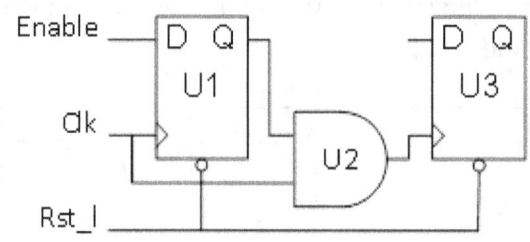

Figure 24: Clock Enable without Clock Management

A timing diagram of the potential for a runt pulse is shown in Figure 25. This diagram shows the Enable signal being clocked into U1 of Figure 24. U1.Q is shown delayed by T_{CK2Q} and it is the input to the AND gate to enable or disable the Clk signal which is the other input to the AND gate. Because of the delay, the output of the AND gate is U3.CK is reduced in length by T_{CK2Q} when Enable goes high. When Enable goes low, the last pulse of U3.CK is only T_{CK2Q} long. In both cases, assuming the clock is a 50% duty cycle and is running at F_{MAX}, the runt pulses can cause timing violations because the high state of the clock would be less than one-half of $1/F_{MAX}$ (the clock period).

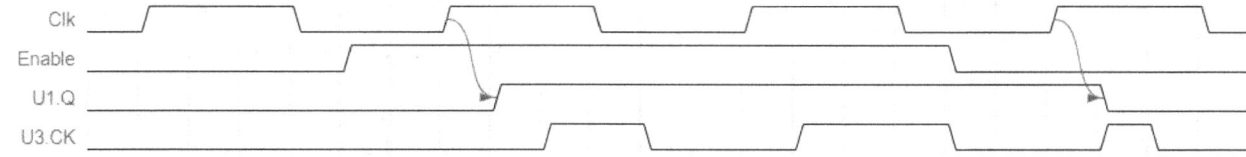

Figure 25: Clock Enable - Runt Pulses

1.9 Timing Closure

It would be a nice day if we designed our logic, synthesized it, and ran it through the place and route tool, and the result of static timing analysis is "Timing Closure". If this happens, it is likely because of one of the three examples below.

- You are a brilliant designer. You thought of every possible case for a timing violation and made the proper design decision to mitigate it.
- You are not pushing the limits of the FPGA you selected.
- You forgot one or more of the timing constraints explained so far, and the static timing analyzer did not have a valid basis for analysis. It probably warned you about this.

Although we all like to think we are brilliant designers, the first item is unlikely. You may not be pushing the limits of the technology, but this might be an opportunity to select a slower speed grade or revert to a technology that is less capable (and cheaper). Finally, recheck the constraints and the warning file that the timing file that the analyzer produces. Make sure all clocks have a frequency constraint, all inputs have a T_{SU} and T_H constraint, and all outputs have a T_{CK2Q} constraint. Bidirectional signals are constrained as both inputs and outputs.

The reality is that there will be timing violations. There are easy and hard ways to fix them. Simply rerunning place and route with a faster speed grade might help. Be sure the faster speed grade part is available, and the price increase is within your budget. Also, if the violations are substantial or if there are a lot of them, you may consider going to a faster, more state-of-the-art technology. This also has availability and cost implications. Finally, you can investigate options for modifying your logic design to reduce the T_P within a timing path. Some examples are shown below, but your actual logic requirements will dictate what techniques you use to improve timing paths.

Pipelining is a technique where latency is added to the design, but the delay is decreased between states allowing the frequency (throughput) to be increased. The schematics in Figure 26 show the changes needed to achieve timing closure using pipelining.

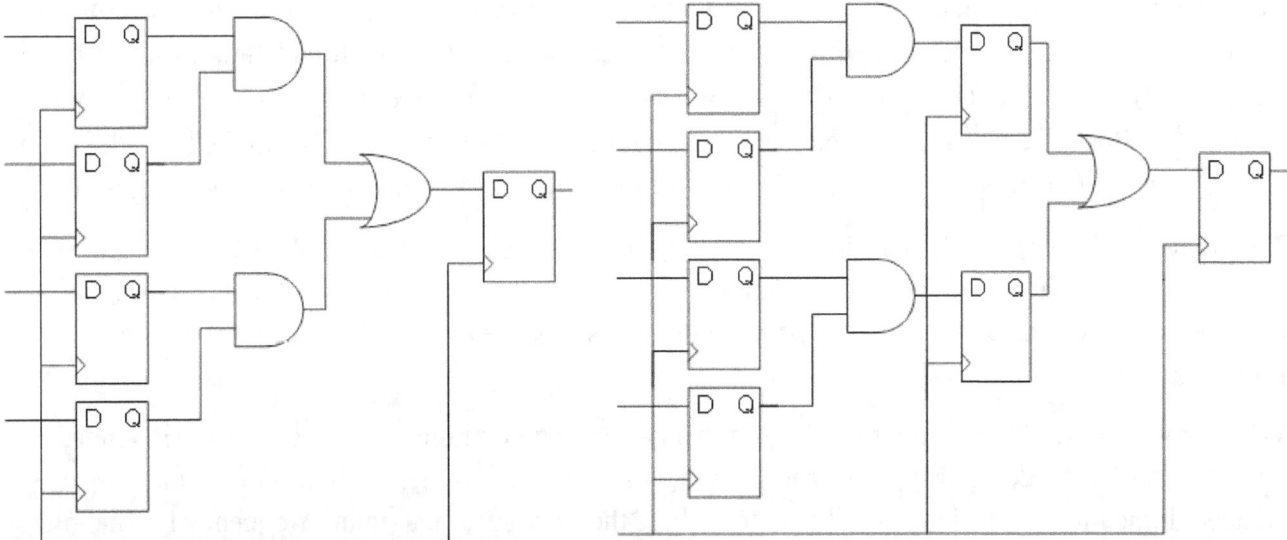

Figure 26: Timing Closing using Pipelining

Pipelining Example:
Given the following parameters for Figure 26, calculate the maximum frequency without and with pipelining and verify that it meets the desired F_{MAX}.

T_{CK2Q} = 1 nS
T_P = 1 nS (per gate level)
T_{SU} = 1 nS
Desired F_{MAX}= 300 MHz

Without Pipelining:

$$F_{MAX} = \frac{1}{(2\,nS + 1\,nS + 1\,nS)} = 250\,MHz$$

Because 250 MHz < 300 MHz, timing closure is not achieved (Timing violation).
The latency equals two clock cycles which is approximately 6.67 nS.

With Pipelining;

33

$$F_{MAX} = \frac{1}{(1\,nS + 1\,nS + 1\,nS)} = 333\,MHz$$

In this case, 333 MHz > 300 MHz, so timing closure is achieved.
The latency is equal to three clock cycles which adds up to 10 nS.

Parallelism is adding multiple copies of logic to reduce delay in other parts of the design. An example of this is shown in Figure 27. Looking at lines 26 and 27 on the left listing, the second assignment uses the index value plus one. This line of code shows up as a timing violation, not because of the lookup but the combinational logic associated with adding one to the index. A possible fix for this is on the right-side listing. Looking at line 10, there is another counter (countp1) declared which will be the count value plus one. It initialized to 3'b001, which is one more than the initialization value for the count. On line 29 (which is line 26 on the right side listing), the index for selector_b is now countp1. Both of these listings produce the same functional results. The code on the left side has an extra counter but no adder in the index. Because of this, the listing on the right side has a higher F_{MAX}.

Techniques for adding parallelism are very application dependent. They generally cause the gate count to go up but they have a higher F_{MAX}. Often these solutions to timing closure problems require the designer to think of creative solutions that are not necessarily the most logic (or power) efficient implementations.

When looking at timing closure, there is always an absolute maximum for a given die/speed grade combination. If the design is synthesized such that the output of a logic block is a direct input to an adjacent logic block. The F_{MAX}, for that case, will be the theoretical maximum frequency for that die/speed grade combination. Of course, there always has to be some logic between Flip-Flops to do something interesting.

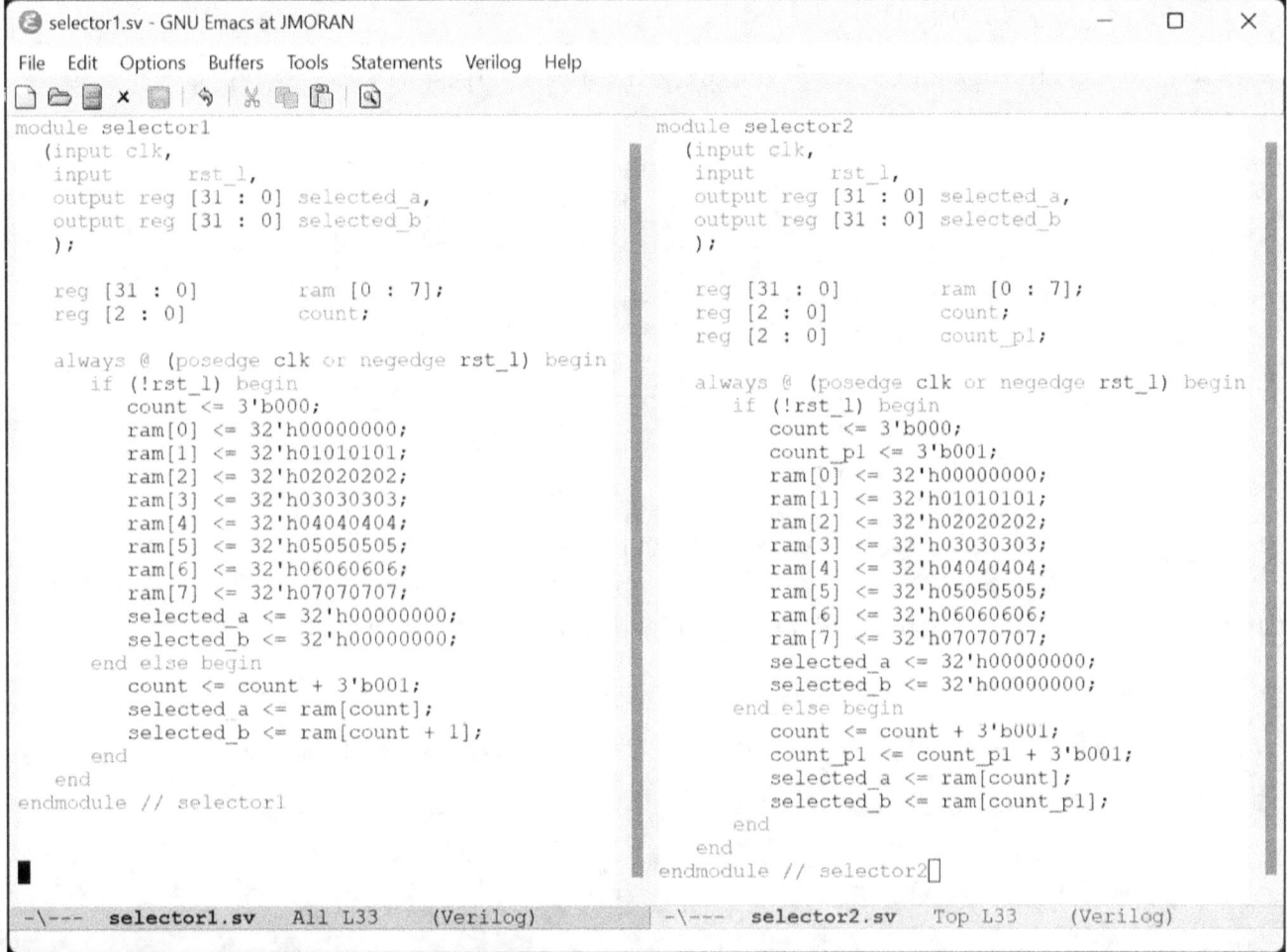

Figure 27: Timing Closure using Parallelism

1.10 Exercises

1) Calculate F_{MAX} using the model shown in Figure 5 using the following parameters.

$T_{CK2P} = 150$ pS; $T_P = 170$ pS; $T_{SU} = 60$ pS; $T_H = 0$ pS.

2) What is the slack of Exercise 1 if the desired clock rate is 2.5 GHz? Is timing closure achieved?

3) Repeat Exercise 1 using the model shown in Figure 7 using $T_{SKEW} = 90$ pS

4) What is the slack of Exercise 3 if the desired clock rate is 2.5 GHz? Is timing closure achieved?

5) Repeat Exercise 1 using the model shown in Figure 9 using $T_{SKEW} = 90$ pS

6) What is the slack of Exercise 5 if the desired clock rate is 2.5 GHz? Is timing closure achieved?

7) What is the maximum allowable skew?

8) If a system has a combination of positive up to 90 pS and negative skew up to 90 pS, what is F_{MAX} of the system?

9) Calculate F_{MAX} for the model in Figure 28 using the following parameters:

All Flip Flops: T_{CK2P} = 150 pS; T_{SU} = 60 pS; T_H = 0 pS.

NAND Gate: Low to High transition, T_P = 90 pS; High to Low transition, T_P = 30 pS;

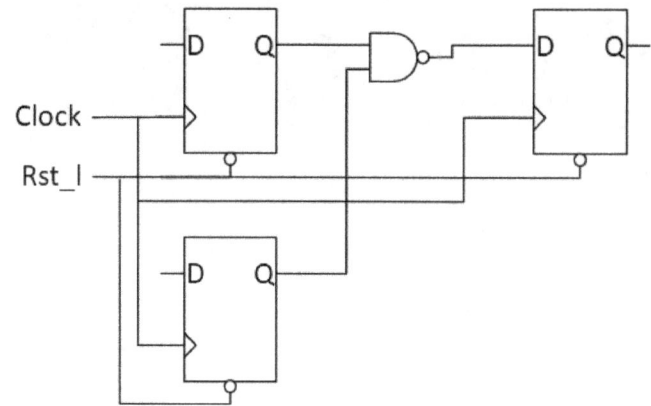

Figure 28: Exercise 2.9 Timing Model

10) Calculate F_{MAX} for the model in Figure 29 using the following parameters:

All Flip Flops: T_{CK2P} = 250 pS; T_{SU} = 300 pS; T_H = 0 pS

Propagation Delays: T_{P1} = 250 pS; T_{P2} = 150 pS; T_{P3} = 100 pS; T_{P4} = 300 pS; T_{P5} = 200 pS.

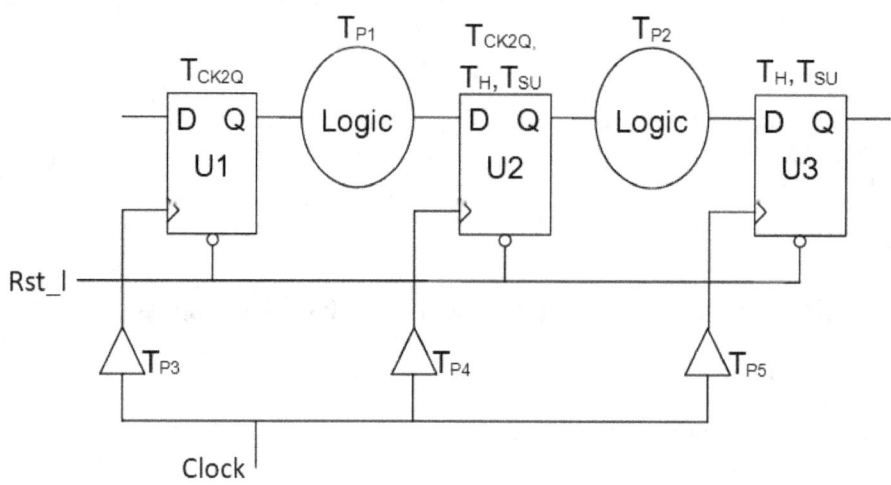

Figure 29: Exercise 2.10 Timing Model

11) Add components to the model in Figure 30 to facilitate a proper clock domain crossing.

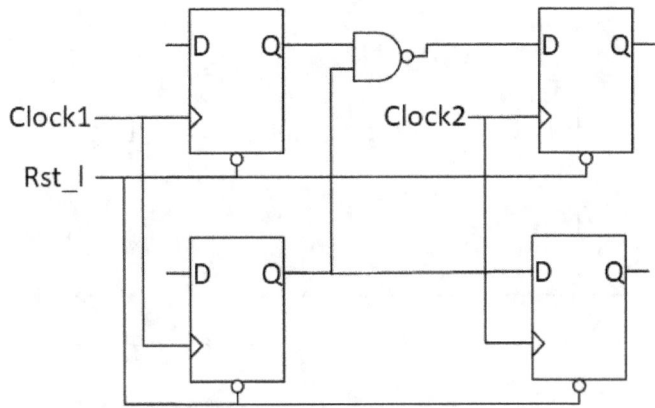

Figure 30: Multi-clock Model

12) Add components to the model in Figure 30 to properly synchronize the Rst_l signal.

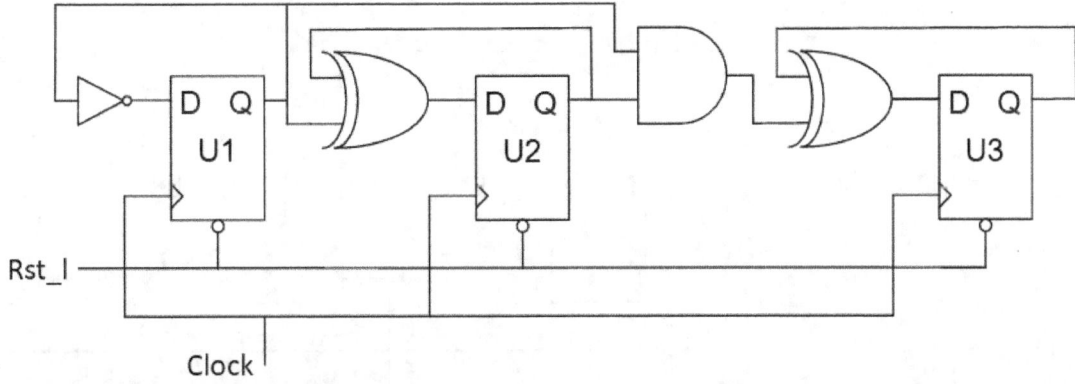

Figure 31: Synchronous Divide by 8 Counter

13) Does the circuit shown in Figure 31 solve the derived clock problem shown in Figure 22 on Page 31?

14) What is F_{MAX} for the model in Figure 32 given the following parameters. What is the latency at F_{MAX}?

All Flip Flops: T_{CK2P} = 150 pS; T_{SU} = 60 pS; T_H = 0 pS.

T_{P1} = 170 pS; T_{P2} = 80 pS; T_{P3} = 100 pS;

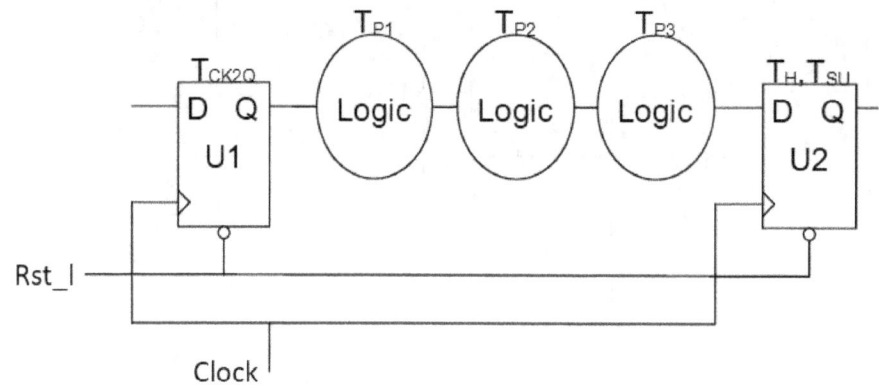

Figure 32: Multi-level Logic Model

15) What is F_{MAX} for the model in Figure 33 using the parameters in Exercise 14. What is the latency at F_{MAX}?

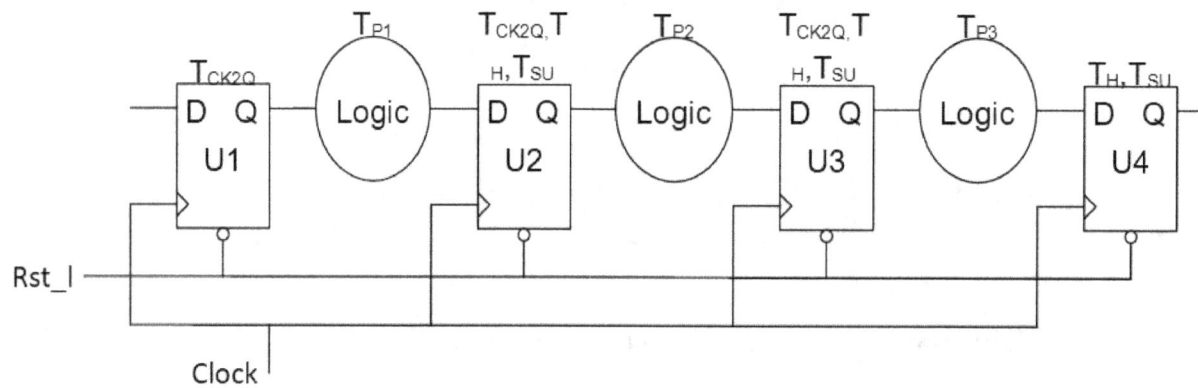

Figure 33: Pipelined Model

16) Draw a model that is a compromise between the multi-level logic model shown in Figure 32 and the pipelined logic model shown in Figure 33 that has an F_{MAX} of at least 2.5 GHz and has the minimum possible latency.

2. Spatial Constraints

As the name implies, Spatial Constraints select the component and how it is connected to the rest of the system.

2.1 Die Selection

Selecting the die to target the design already assumes the manufacturer and family have been selected. Manufacturer selection usually goes beyond the scope of the design and into potential existing relationships with a manufacturer, experience with design tools, unique features of the family, and, of course, cost. Family selection comes from the general size and performance requirements of your application.

The actual die selection is fundamentally based on the resources on the die, the number of logic cells, the amount of memory, and the number of Input/Output (IO) blocks. There is a strong relationship between the selection of the die and the selection of the package.

Two other components of die selection are the speed grade and temperature range. The speed grade is commonly based on processing and binning by the manufacturer. Parts on the fast side of the processing curve are processed in a faster technology and are generally sold as faster speed grades. These faster parts will have an easier time achieving timing closure. They will also cost more. The temperature range defines the range of ambient temperatures where the component will run correctly. Because temperature is a strong component in the variation of timing parameters, timing closure will be more difficult over a wider temperature range. Not all die are available in every speed grade and temperature range. Sometimes the same part is constrained at different temperature ranges, which yields different timing results.

2.2 Package Selection

Given a selected die, there is a limited number of available packages. If possible, the actual package selected should support more than one die. This allows the designer to migrate to a larger die if more resources are required, or a smaller die to reduce cost. It is common to use the largest die available for a given package during prototyping and switch to a smaller die in production. An example of a die/package selection table is shown in Table 2.

An example showing the implication of selecting a package from Table 2 would be the selection of the 384 FBGA package. If this package was selected, the designer would have the option of using a 100,000, 200,000, or 500,000 gate die. This is a lot of flexibility to make trade-offs between cost and features. There is one caveat. Notice that in Table 2 the number of IO for the 384 FBGA is 240 using the 100,000 gate part and 260 for the 200,000 and 500,000 gate part. If more than 240 IOs are required, the selection is limited to 200,000 or 500,000 gate dies. Also, if the pin selection was done such that the unavailable pins on the 100,000 gate die were used, the designer would be limited to the 200,000 and 500,000 gate parts. For this reason, the designer should watch out for this caveat and make pin selections accordingly.

Table 2: Package/Die Selection Table

Package	Die Size (Gates)						
	10,000	20,000	50,000	100,000	200,000	500,000	1,000,000
64 CSP	32	32					
208 QFP		112	112	112			
256 BGA		112	112	112			
384 FBGA				240	260	260	
1529 FBGA						1000	1000

2.3 Pin Selection

Pin selection is a complicated task in FPGAs that are in large packages. The package shown in Figure 34 is a very large fine-pitch ball grid array. Connecting the internal balls is difficult. To make things more manageable, the manufacturers inserted the power and ground pins in the center of the chip. The pins in the center are easier to connect to internal planes. In the package shown in Figure 34 the power and ground pins are shown as (+) and (-), respectively. This is a 16 by 16 array representing a total of 256 pins. This leaves 1344 pins to be used as signals for other functions. The critical thing to note is that the remaining pins form twelve concentric squares around the central power and ground connections.

Also, as shown at the bottom of Figure 34, only one signal fits between the columns of balls. Figure 35 shows typical dimensions for a PCB trace. In this case, the board is constructed with 4 mil traces and 4 mil spacing. The ball connections are 8 mils in diameter. This is a required construction for a Fine-pitch Ball Grid Array (FBGA) with 50mm (20 mil) ball spacing. Therefore, fully routing this part will require a Printed Circuit Board (PCB) with twelve signal layers. The corners have unique routing problems and usually have no-connect pins or special pins. Because PCBs generally need about one power or ground layer for every two signal layers and there must be an even number of layers. This adds up to an 18-layer board. A stack-up of an 18-layer board is shown in Figure 36. A PCB with 18 layers the fine traces and spacing is very expensive. FPGAs that support 1600 ball packages are also costly.

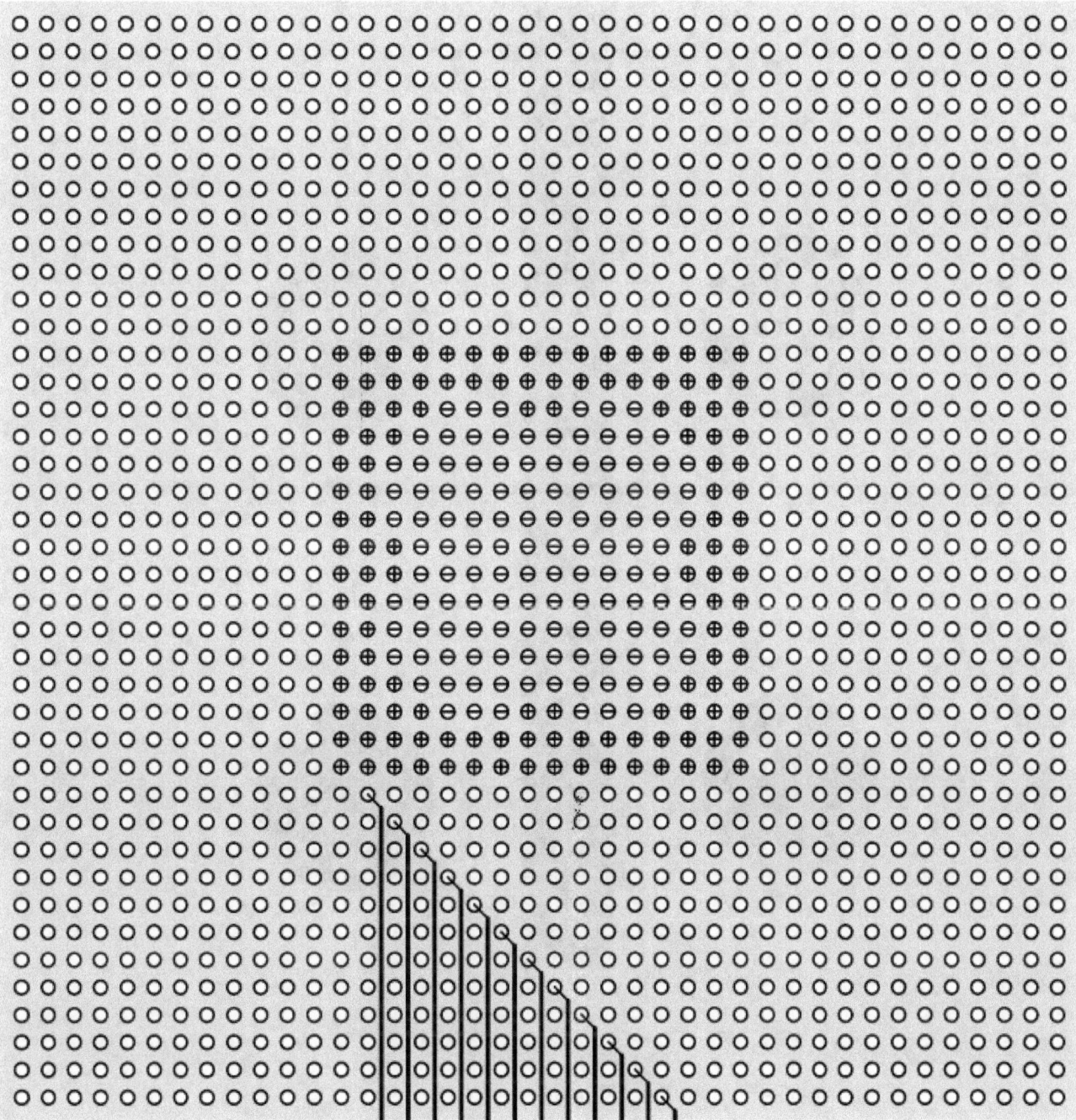

Figure 34: 1600 Ball Package Routing

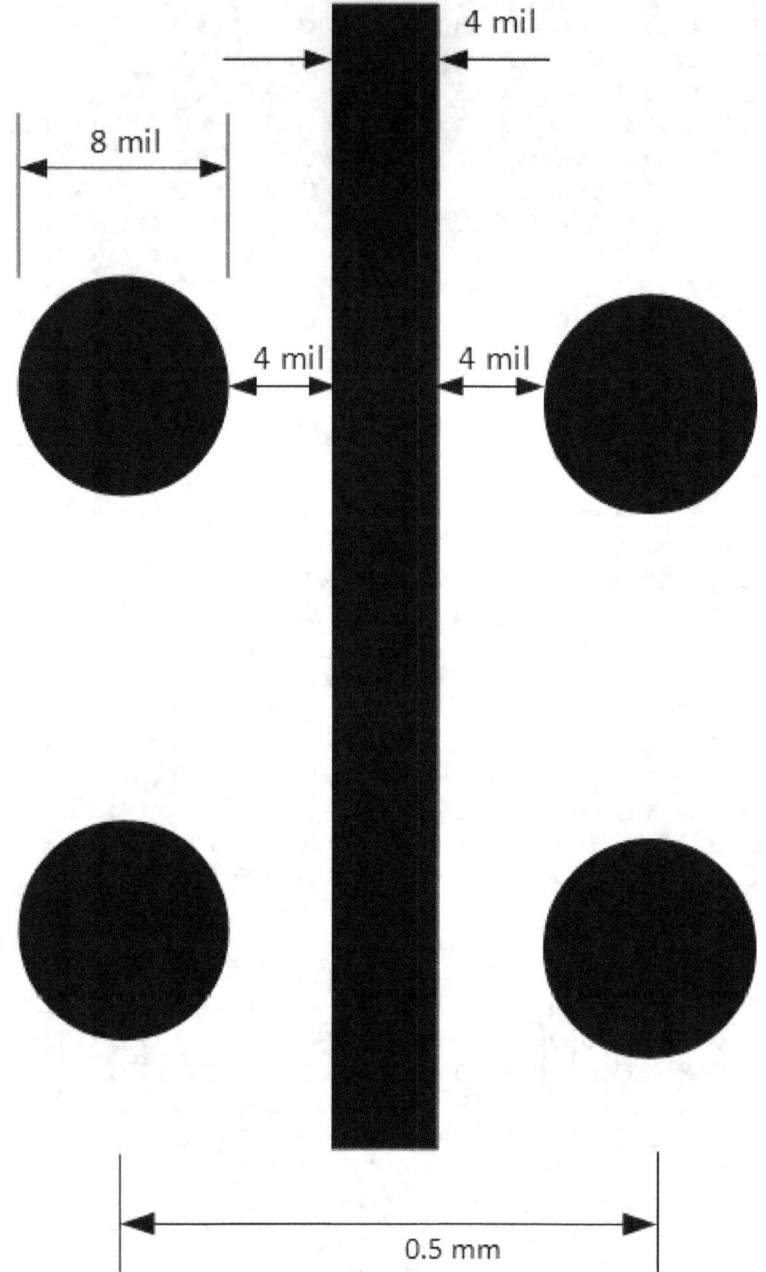

Figure 35: Fine Routing Example

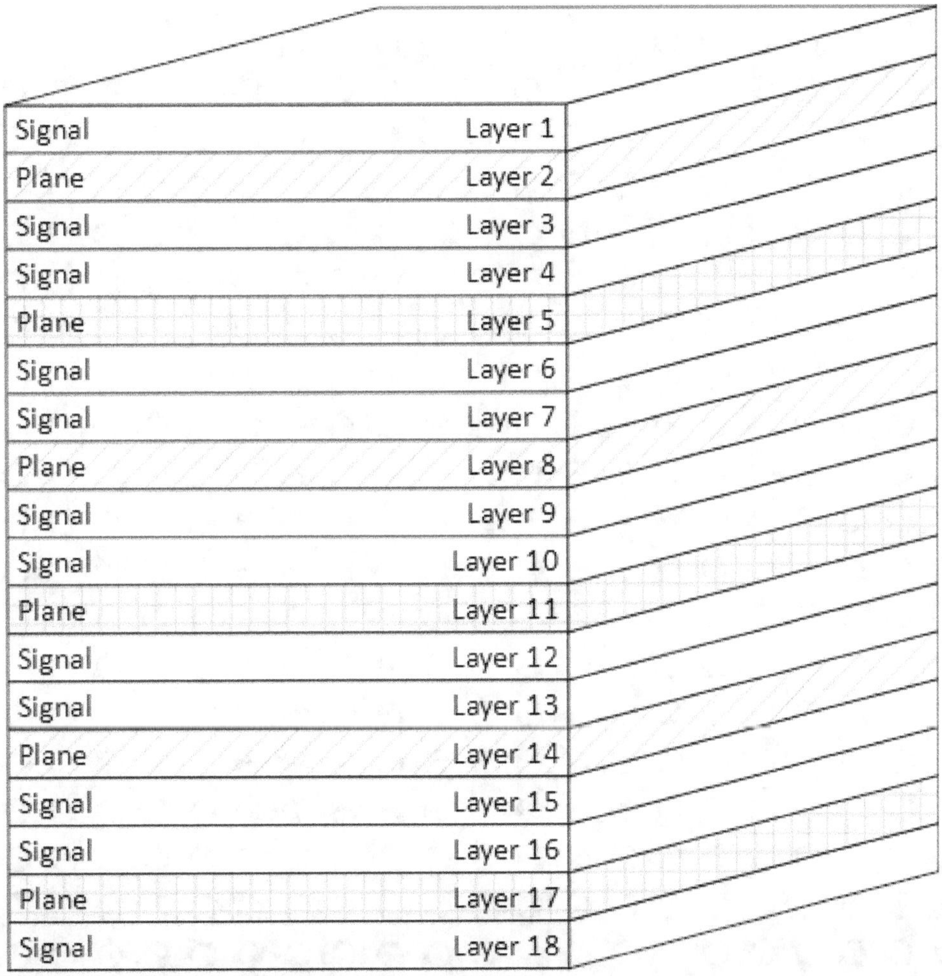

Signal	Layer 1
Plane	Layer 2
Signal	Layer 3
Signal	Layer 4
Plane	Layer 5
Signal	Layer 6
Signal	Layer 7
Plane	Layer 8
Signal	Layer 9
Signal	Layer 10
Plane	Layer 11
Signal	Layer 12
Signal	Layer 13
Plane	Layer 14
Signal	Layer 15
Signal	Layer 16
Plane	Layer 17
Signal	Layer 18

Figure 36: 18 Layer Stackup

Smaller devices are less of a problem but must be carefully handled. Figure 37 shows a 576 package. This package is 24 x 24 with 0.8mm (32 mil) spacing. Because of the wider ball spacing, the total size of the package is about the same as the 1600-pin package shown above. The advantage of the smaller package is that two traces can be brought through a column of balls, as shown in Figure 38. With a 10 x 10 matrix of power and ground balls for a total of 100 pins. This leaves 476 pins to use as signals, and there are seven concentric squares. Connecting seven rows of pins with two traces between balls takes 4 layers. Applying the principle of one power or ground plane per two signal layers yields a stack-up with only 6 layers shown in Figure 39.

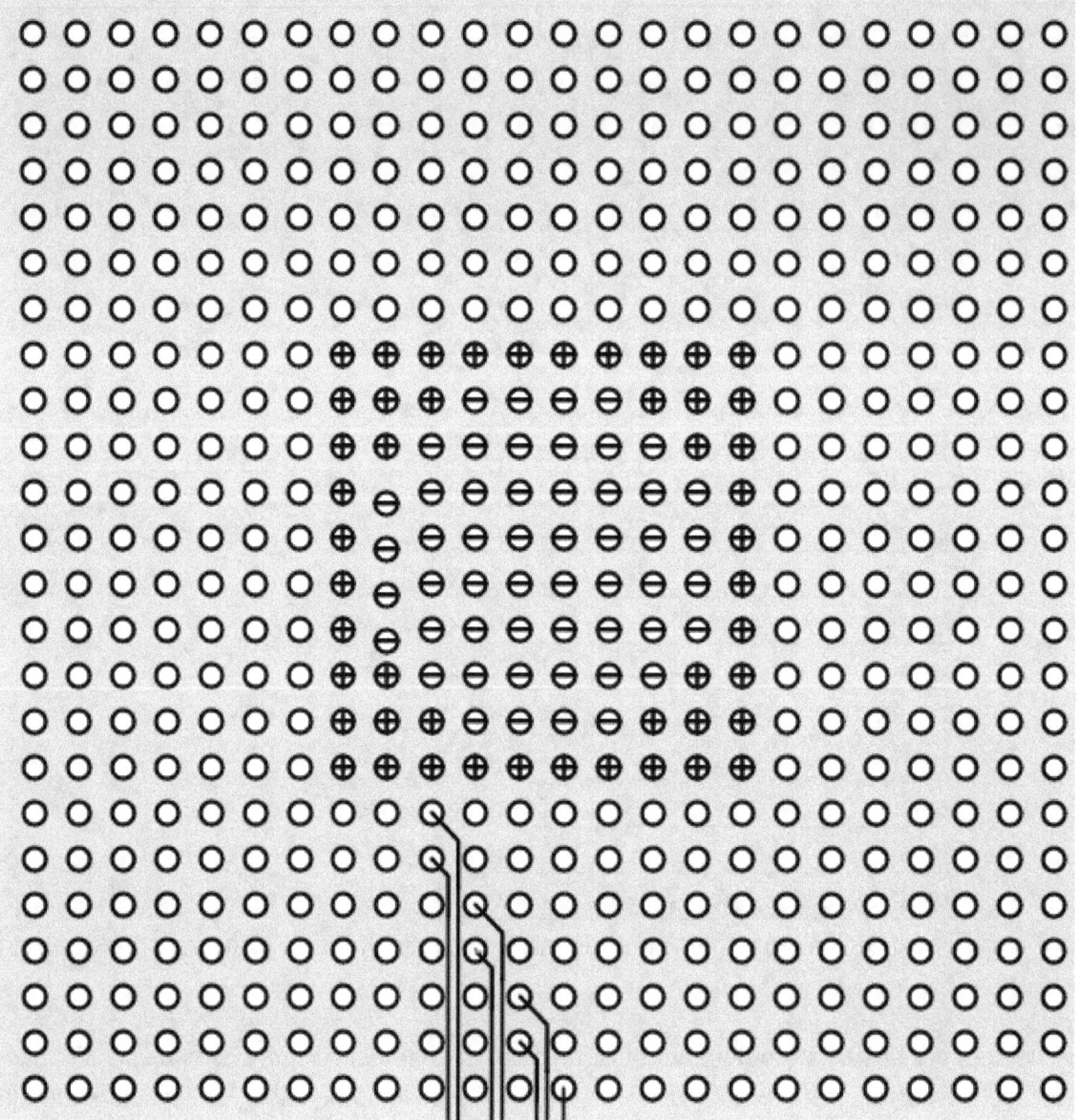

Figure 37: 576 Ball Package Routing

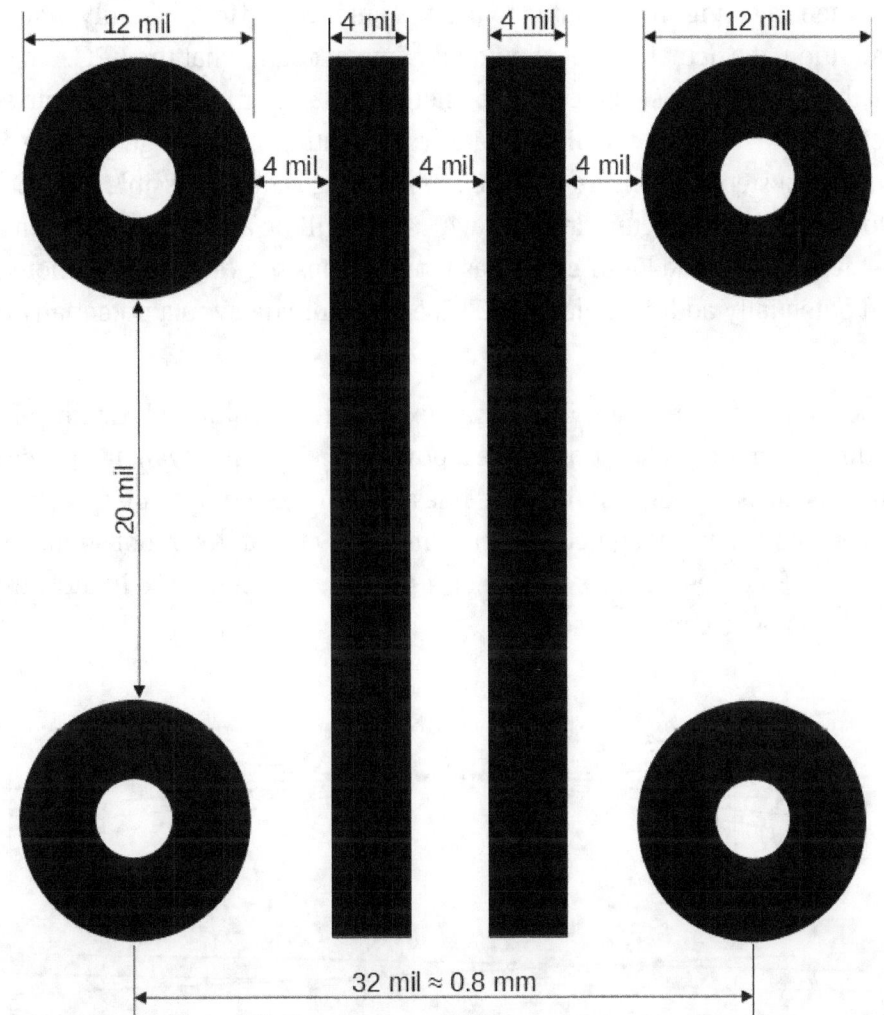

Figure 38: Coarse Routing Example

Signal	Layer 1
Plane	Layer 2
Signal	Layer 3
Signal	Layer 4
Plane	Layer 5
Signal	Layer 6

Figure 39: 6 Layer Stackup

The obvious rule when selecting the pinout is to use the outer rows first. Usually, only a subset of the pins is used. In addition, the actual pinout should take into account what the FPGA is connecting to. One possibility is that this FPGA is connecting to another FPGA. In this case, the pin selection is flexible on both sides. By selecting the pins to satisfy the routing rules for getting the pins escaped from the device, the next step is to ensure that the routing is as easy as possible. If a poor job is done, the routing will look like the top picture in Figure 40. This will be a complex problem for the PCB designer to solve. It will require a lot of extra vias (connections between layers), more holes in the power planes, and potentially additional layers. These add more delay to a potentially difficult timing closure problem.

Conversely, if a meticulous pin selection is done, an exact parallel solution is achieved. This will give the best possible timing solution, but it does have a downside. The disadvantage of the exact parallel solution is that there is crosstalk, especially if the lines are long. A compromise is shown in the center of Figure 40. This example has enough crossover to mitigate crosstalk but not so much that it will cause difficulties in the PCB design. It is not the best possible timing, but as in most things in engineering, it is a reasonable compromise.

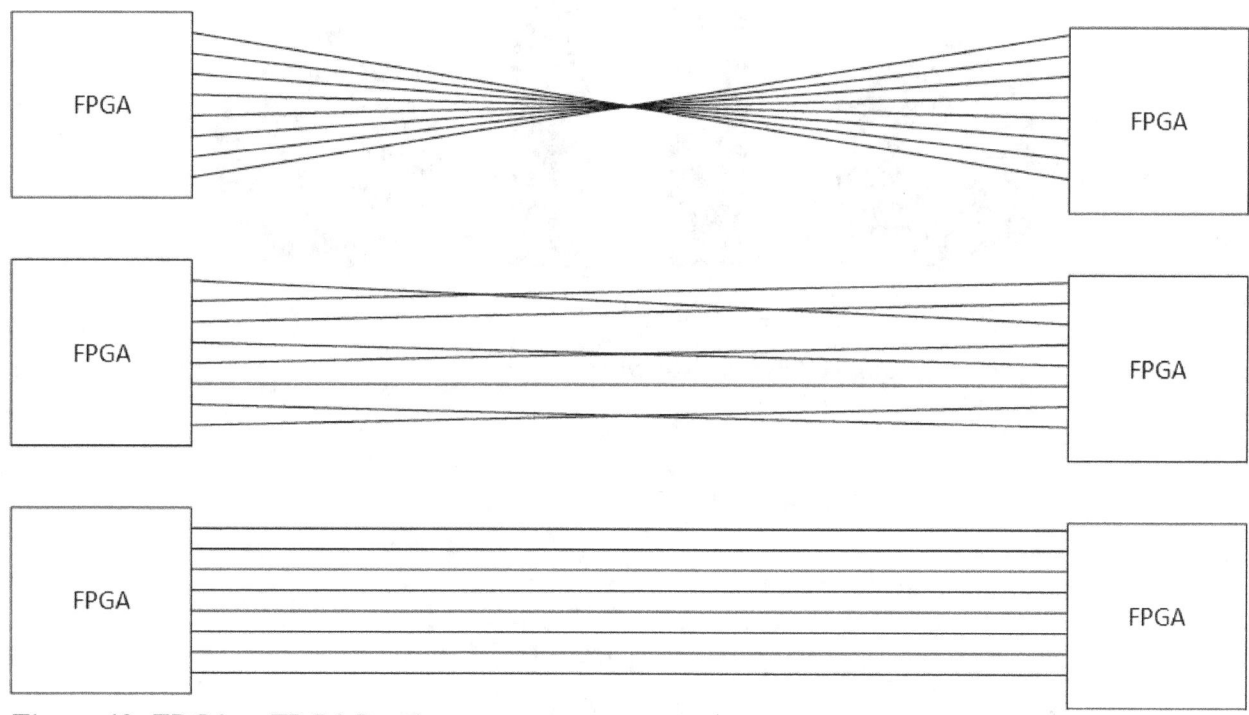

Figure 40: FPGA to FPGA Routing

Figure 41 is essentially the same as Figure 40 but the devices on the right side are fixed pinout devices like processors, memories, and ASICs. In this case, The FPGA on the left side is the only device with a flexible pinout, and more work may need to be expended to achieve optimal pinout. It should be noted that there is some flexibility in the pin selection of an ASIC, but once it is built, the flexibility goes away. The case of a fixed pinout device connected to another fixed pinout device is not covered in this text but is a complex problem to solve for PCB designers.

46

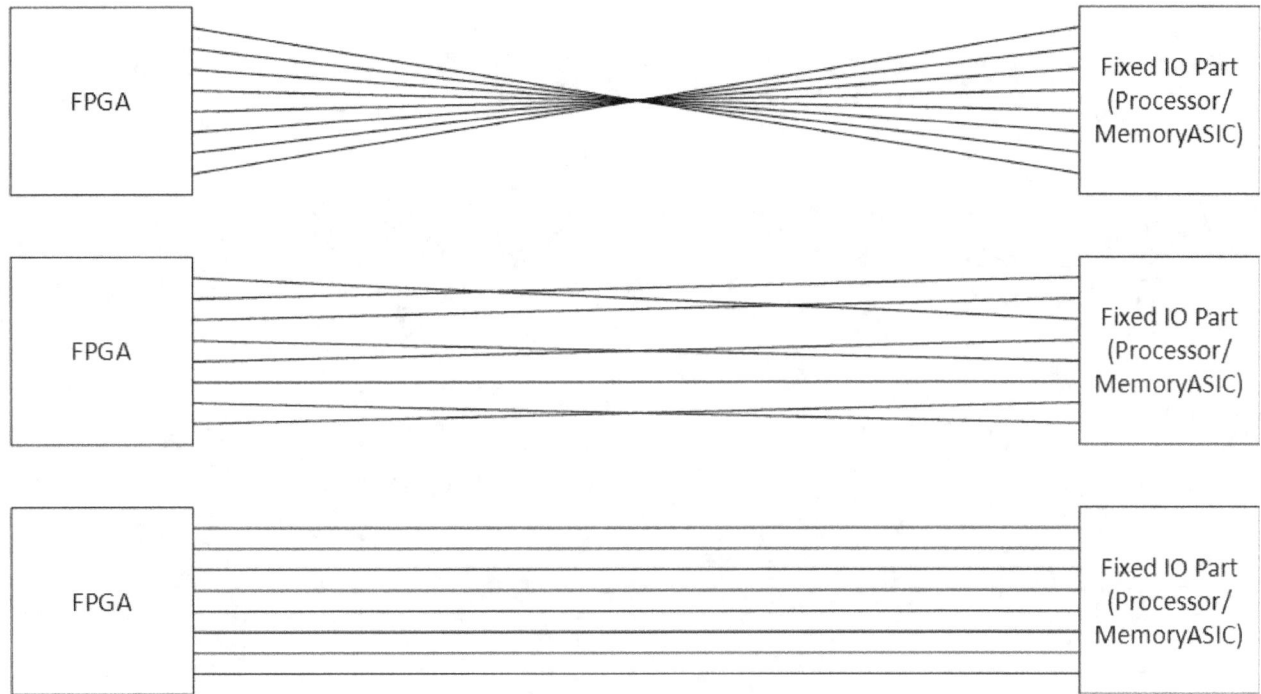

Figure 41: FPGA to Fixed Pinout Device

Another thing to consider is how the routing is handled when connecting to more than one external chip. The case shown in Figure 42 has two external buses on the input, A and B. On the left side of the figure, A and B are intertwined on the input. If A and B go to different devices on the PCB, all signals will cross, making for a difficult PCB design. In a practical system, it is not out of the question for these buses to be much larger, 32 or 64 bits. On the right side of Figure 42, the buses are grouped, so each bus is routed to its respective device without crossover. The consequence of this is that the crossovers are inside the FPGA. Putting the crossover inside the FPGA is better because the routing inside the FPGA is much shorter and faster than the board routing and has less impact on timing closure.

FPGA IOs are in banks with assigned power supplies. Each bank can have its own power supply. In a case where all the banks have the same power supply (V_{CCIO}), there is not much to worry about. If there is more than one different power supply, the banks must be selected such that the banks with like power supplies are adjacent. The left side of Figure 43 shows a case where this is done correctly. In this case the two bank power supplies could be on the same layer with a split between them. The bank selection shown on the right is not laid out correctly. In this case, the power planes for the two banks would most likely need to be laid out on separate layers because of the cross in the middle. If there are more (3, 4, 5…) power supplies the layer count would start to increase, and the cost of the printed circuit board would also increase.

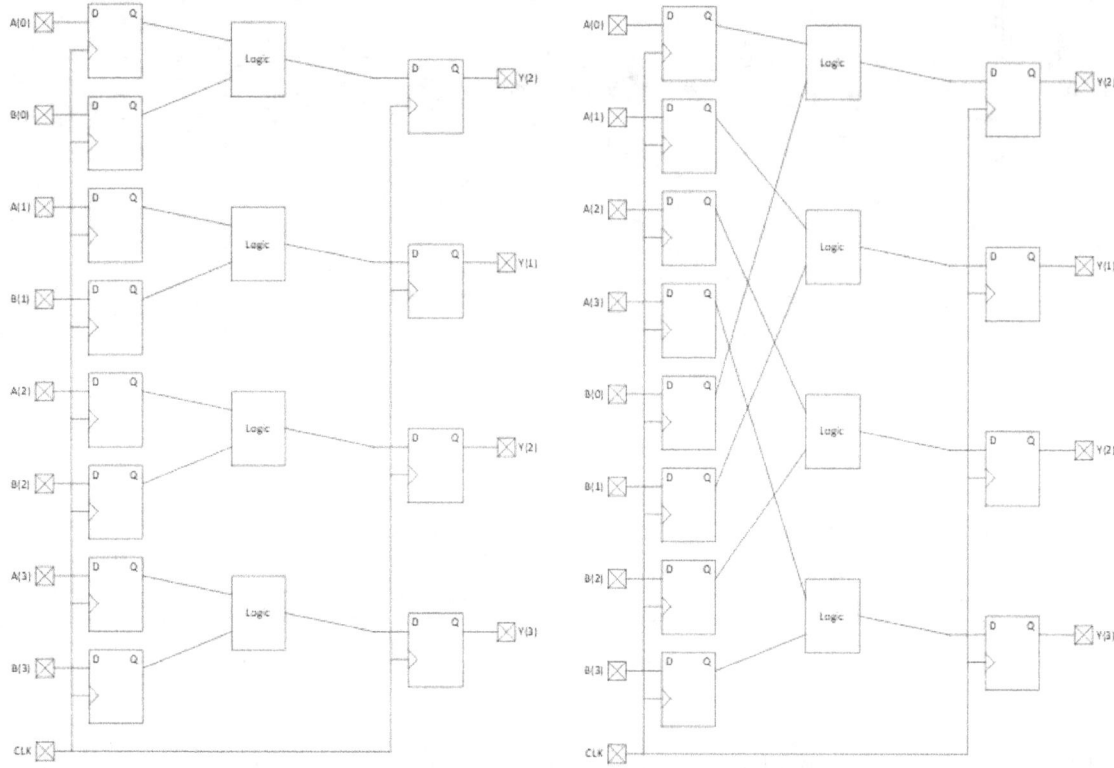

Figure 42: On-chip vs. Off-chip routing

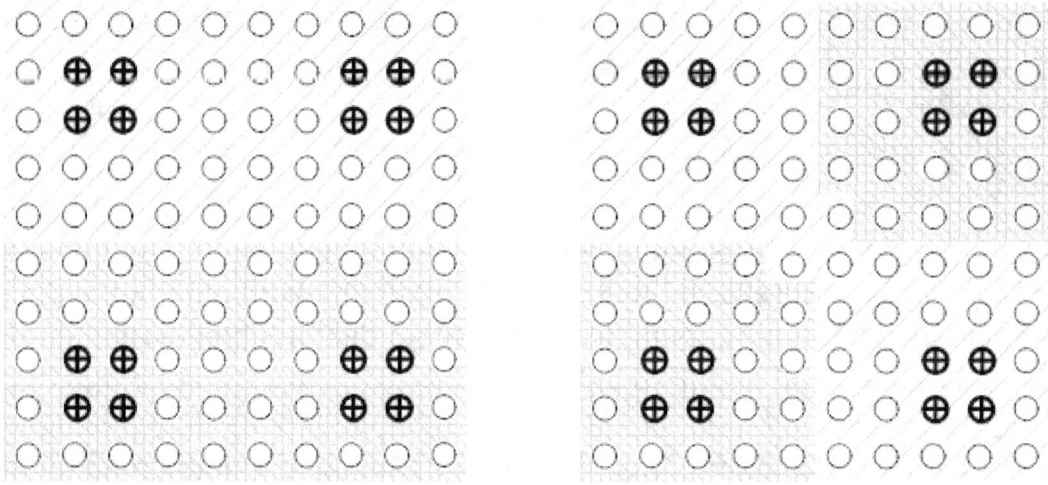

Figure 43: Power Bank Selection

When selecting signals in a bank, simultaneously switching outputs should be evaluated. In synchronous designs, having the signals on a bus change as close to the clock edge as possible is advantageous. On the IO of a chip, simultaneous switching can cause an instantaneous current draw that can cause a phenomenon called ground bounce. Ground bounce is caused by inductance in the ground plane, which turns resistive at high frequencies. The resistance causes a voltage drop that changes the ground reference for the input signals. To reduce the effect of simultaneous switching:

- The output signals are spread out among multiple banks, which makes them harder to route.
- The drive strength, described in Section 2.5, of the signals is reduced, which lengthens T_P and reduces F_{MAX}.
- Small random delays can be added to the signals, so they are not changing simultaneously (the manufacturer specifies what is considered simultaneous switching)

The above options each have a downside but may be necessary if a potential simultaneous switching output problem exists.

There are many other constraints to be considered in PCB design. An FPGA designer must work closely with the PCB designers at every step. FPGA manufacturers publish guidelines for the use of their parts on PCBs. These guidelines must be vetted by the FPGA designer to prioritize which guidelines are most important for the given application. This information should be communicated to the PCB designer, and finally, the FPGA designer should verify the final PCB design before it is fabricated. Some FPGA designers simply hand the PCB design guidelines to the PCB designer. This is ineffective because there are decisions to be made that are based on the specific configuration of the FPGA. For example, the PCB designer does not know if a given signal is differential or single-ended unless the FPGA designer tells them. To further improve this process, signal integrity simulation tools, whether static or dynamic, should be utilized where available.

The technique for designing PCBs is unique for FPGAs. Fixed devices, like ASICs, microprocessors, or memories, have known pin locations, signal types, and power requirements are found in their datasheet. These parameters are not fixed in an FPGA. The FPGA designer must provide information about the specific configuration to the PCB designer. In addition, when it gets down to actual PCB routing the PCB designer may want to swap pins on the FPGA. Care should be taken to only swap pins between banks with the same V_{CCIO} power supplies.

2.4 Input-Output Voltage Selection

An additional challenge in constraining the Input-Output (IO) connections is the IO voltage. This is selected based on the requirements of the device that you are connecting to. If this is another FPGA, the problem is easier because you have control over both sides. If the FPGA is connected to an ASIC, microprocessor, or memory, the IO voltage must be selected to match. This seems like an easy problem to solve, just set the voltage to the correct voltage and it will connect. There are two limitations to setting the IO voltage. First, when the IO voltage constraint is set on a bank, all of the connections on that bank are required to comply with that constraint. Also, every time a new IO voltage is used, a power supply with that voltage must be included. If too many different voltages are used, the additional components can add significant space and cost to the system. To mitigate the effects of both of the additional components, work to reduce the number of power supplies is necessary. A first-level analysis would include a study of the IO voltage options of the external devices. These devices occasionally have limited options, but it is unlikely they will solve the whole problem.

To properly analyze the voltage options, we first need to define some terms associated with IO connections, commonly referred to as DC characteristics. In the worst-case, incompatible IO voltages may require external level-translators. Adding external level-translators is the last resort because of the space, power, delay, and cost penalties involved.

V_{CC} – Core voltage of the FPGA, specified as a nominal voltage and a range (usually ±5%).

V_{CCIO} – IO Voltage – This is the minimum voltage applied to the IO voltage supply pins. There is usually an absolute maximum on these pins to avoid damage. This is set as a spacial constraint.

V_{IL} – Voltage Input Low – This is the maximum voltage that will be interpreted as a logic 0 by the input connection. There is also an absolute minimum associated with V_{IL}.

V_{IH} – Voltage Input High – This is the minimum voltage that will be interpreted as a logic 1 by the input connection. There is also an absolute maximum associated with V_{IH}.

V_{OL} – Voltage Output Low – This is the maximum voltage driven by an output connection to signal a logic 0. V_{OL} must be lower than V_{IL}.

V_{OH} – Voltage Output High – This is the minimum voltage driven by an output connection to signal a logic 1. V_{OH} must be higher than V_{IH}.

Now that the IO parameters are defined, we can describe a phenomenon known as noise margin. The high noise margin, shown in Figure 44 is defined as $V_{OH} - V_{IH}$. This margin accounts for power supply variations and noise coupled with chip-to-chip signaling. The low noise margin, shown in Figure 44, is defined as $V_{IL} - V_{OL}$. This margin accounts for ground noise and noise that is coupled on the chip-to-chip-signaling. These margins can also be used to evaluate IO signals between devices where the IO voltages differ. It should be noted that the low parameters are usually not dependent on V_{CCIO} because they are referenced to ground, but high parameters are generally referenced to V_{CCIO}.

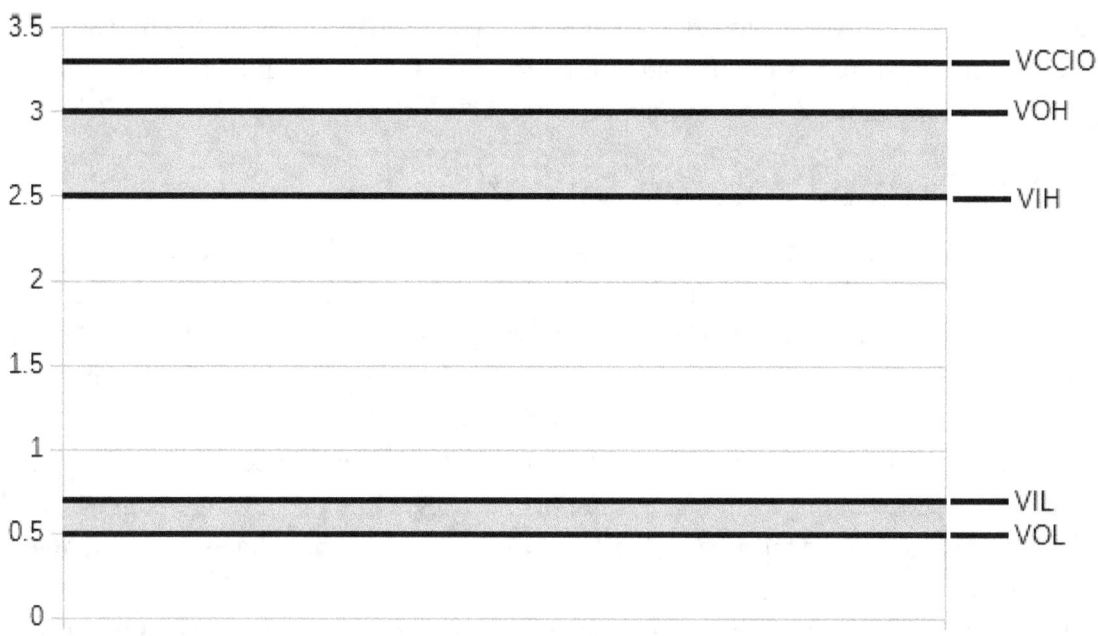

Figure 44: Noise Margins

<u>Noise Margin Example</u>

Calculate the noise margin for a connection between two devices as shown in Figure 45 given the following parameters:

V_{CCIO} = 3.3V+/- 5%
V_{IL} Max = 0.7,$_v$ Min = 0
V_{IH} Min = 1.8V, Max = V_{CCIO} + 0.5V
V_{OL} Max = 0.5V
V_{OH} Min = V_{CCIO} – 0.5V

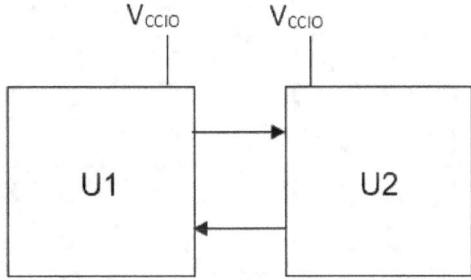

Figure 45: Noise Margin Example

Low Noise Margin   = V_{IL} – V_{OL} = 0.7V – 0.5V = 0.2V
High Noise Margin   = V_{OH} – V_{IH} = V_{CCIO} - 0.5V – 1.8V
                         = 3.3V*.95 – 0.5V – 1.8V (Use the low range of the power supply tolerance)
                         = 3.135V – 0.5V – 1.8V
                         = 0.835V

2.5 Output Load Constraints

The load that the output of a given input pin of an FPGA will need to drive is an important constraint. This constraint is derived from the system design and includes the sum of the capacitance of the printed circuit board trace plus the sum input capacitance of each of the components that the pin will drive. The total capacitance, usually stated in picoFarads (pF), is set as a constraint for each output pin. This value has an effect on the output pin's T_{CK2Q} so it should be thought of as a temporal constraint masquerading as a spacial constraint. The output load constraint has an effect on the power calculations done by the implementation tool. FPGA implementation is discussed further in Chapter 4.

In addition to setting load constraints on outputs, some FPGAs allow the designer to set the drive strength of each output pin. Setting a higher drive strength will lower that pins T_{CK2Q} for a given load. The downside of setting a higher drive strength is higher power and potentially more switching noise. A good rule of thumb to use is to set the lowest possible drive strength that will achieve board level timing closure. The drive strength constraint is usually stated in milliamps (mA) and is available in values that are powers of two (1 mA, 2, mA, 4 mA …).

2.6 Exercises

1) Select an FPGA package die combination from a major manufacturer with at least 100,000 Flip-Flops and at least 256 Signal Pins. Be sure that a development board is readily available for this component.

2) Estimate the layer count including power and ground for the package shown in Figure 20 assuming bank E is fully connected and all other banks are unconnected.

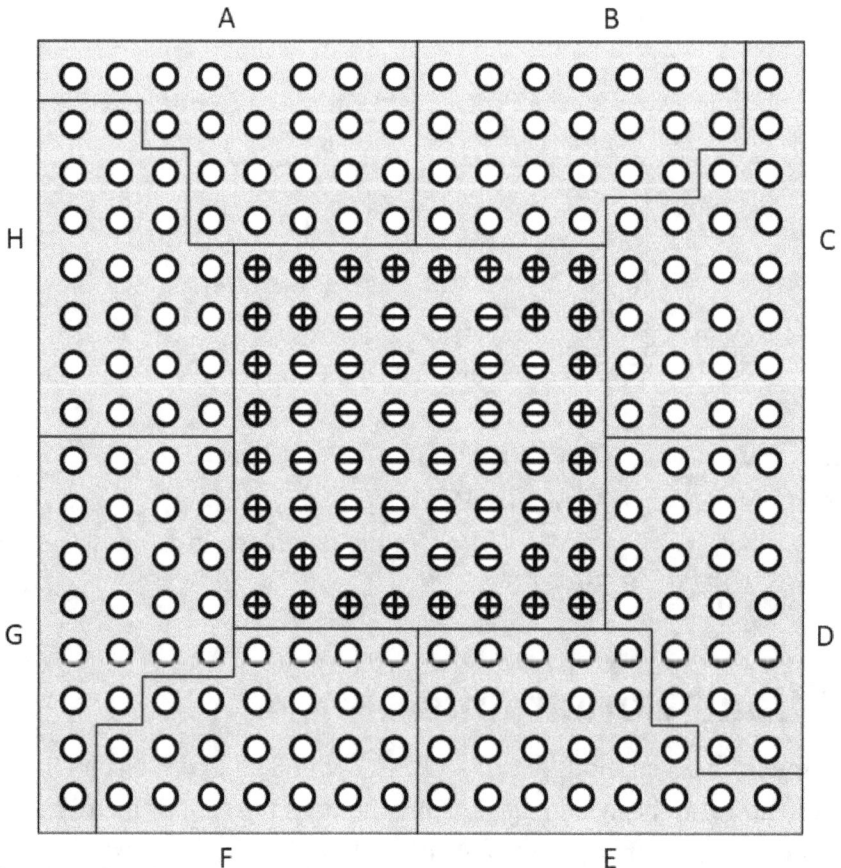

Figure 46: Bank Routing

3) Estimate the layer count including power and ground for the package shown in Figure 20 assuming bank E is full and all other banks are fully connected.

4) Using Figure 45, calculate the voltage margins for each direction using the DC characteristics shown in Table 3.

Table 3: DC Characteristics for Exercise 4.4

	U1		U2	
V_{CCIO}	2.5V +/- 5%		3.3V +/- 5%c	
	Min	Max	Min	Max
V_{IL}	-0.2V	0.7V	-0.2V	0.7
V_{IH}	$V_{CCIO} - 0.5V$	$V_{CCIO} + 1.0V$	$V_{CCIO} - 1.5V$	$V_{CCIO} + 1.0V$
V_{OL}	-	0.5V	-	0.5V
V_{OH}	1.8V	V_{CCIO}	2.6V	V_{CCIO}

3.   Case Study: Circuit Design is not System Design

It's not good news when your boss comes into your office and says, "I need you to do something, and you're not going to like it". I had worked with this boss at a previous company. We trusted each other and knew each other's strengths. This was not the first time he made this request. The previous time was a case where a designer had thrown together a design without following good design principles. Unfortunately, we were in the same situation. This was a startup, he was the VP of engineering, and I was the Hardware Architect. Our first product was done and shipping. To accelerate things, one of the second products was being developed by a consultant while I was working on the follow-on to the shipping product. Once the consultant delivered the prototype, it was obvious that I needed to jump in.

The consultant is a very experienced Digital Signal Processor (DSP) designer. By all accounts, his DSP design was perfect. For some reason, he insisted on targeting the DSP design to the FPGAs and designing the PCB. He either did not have the skills to target the FPGA and design the PCB, or he rushed through it. I think it is the former. The design consists of eight FPGAs, four each of two different DSP designs. To protect proprietary information, I've left out the actual functions, and I'm just calling them input FPGAs and output FPGAs. Figure 47 shows the arrangement of the devices. Each of the input FPGAs has four 32-bit buses that go to the four output FPGAs. This is a total of sixteen 32-bit buses all synchronous running at 200 MHz which was near the limit of the technology at the time. A single clock source on the board distributed the 200 MHz clock to all eight FPGAs.

3.1 Power Analysis

When the system was delivered, it was not working. It was possible to load a single input FPGA manually and a single output FPGA put a stimulus into one input port, and see the correct output on the first output port. This brings us to the first problem. None would load when trying to automatically load the FPGAs from their local serial EEPROMs. I found that suppressing the load of seven of the FPGAs allowed the eighth FPGA to program correctly. I suspected there was a power issue because the FPGAs draw a good bit of power when they are being loaded. Because of the method used to load them, it was possible to load them one after the other. When I worked this out with some board rework, all eight were programmed correctly every time. Problem solved, or so I thought.

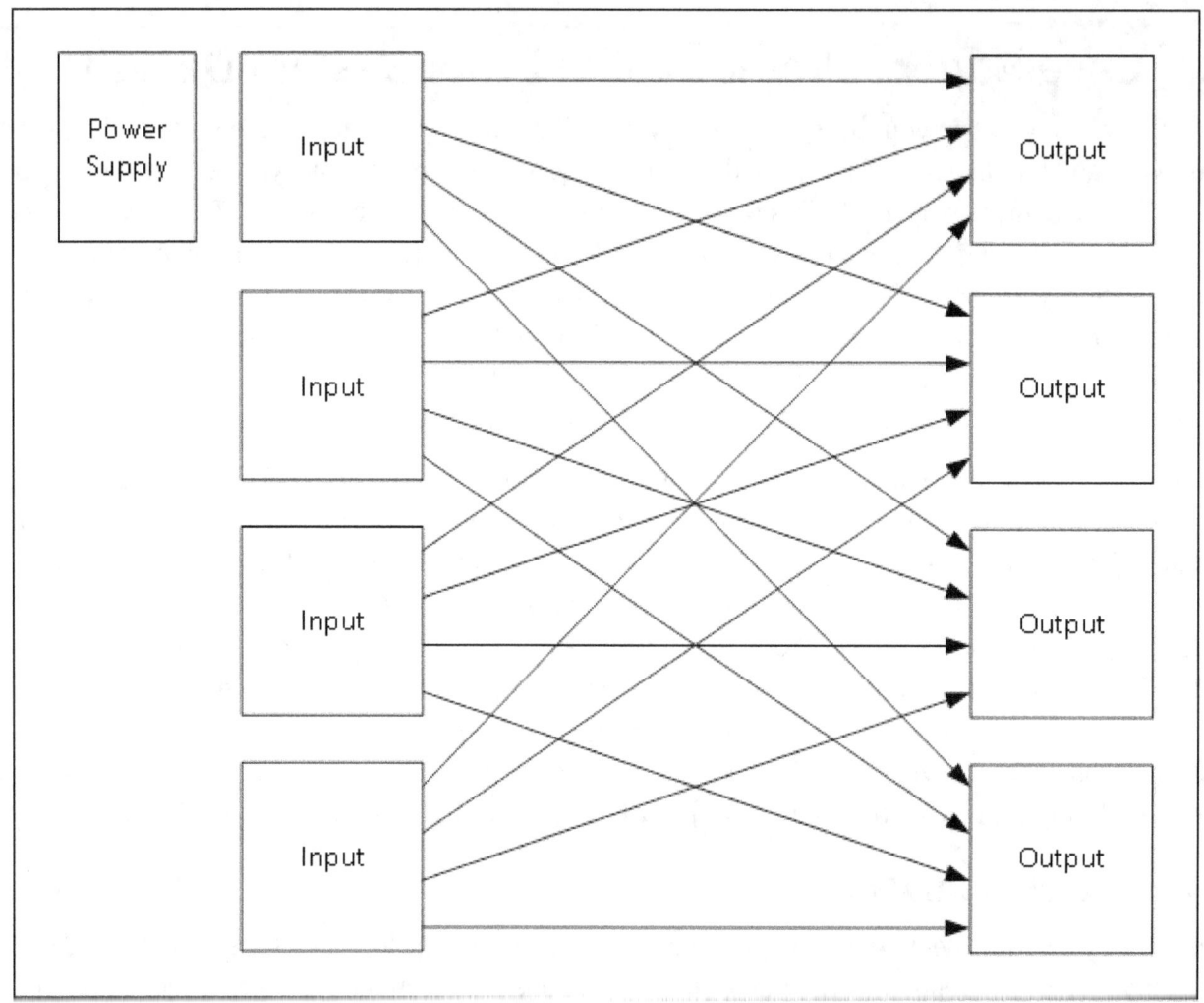

Figure 47: PCB Layout of FPGAs

Now with all FPGAs programmed, I could insert signals on all ports. I knew that the function worked, but when I looked at the FPGAs' output ports, only two had any output, and it was only marginally correct. My reaction in this situation would be to scrutinize the schematic, but I still had the power supply instrumented from the earlier problem. I measured the output of the power supply connected to the core of the FPGAs. It was in regulation with no appreciable noise. I started to make local measurements of the FPGAs, and I found that the supply voltage of some of the FPGAs was too low. This prompted a look at the PCB routing, specifically the voltage plane.

An engineer once told me that experienced engineers sometimes do not catch "rookie mistakes" when reviewing designs because they have not made them in a while. What was staring me in the face while I looked at the PCB artwork was well beyond a rookie mistake. To save board cost, the PCB designer (the consultant) mixed routes between the FPGAs and the voltage plane on the same layer. This caused the power to be routed in a "U" pattern around the perimeter of the FPGAs. In addition, the connection between the power supply output and the plane was insufficient. The result was that as the power traversed across the FPGAs, the small power plane's resistance caused the voltage to drop. Figure 48 shows the routing of the plane. It turns out that the third and fourth output FPGA were the ones that kind of work. This made sense.

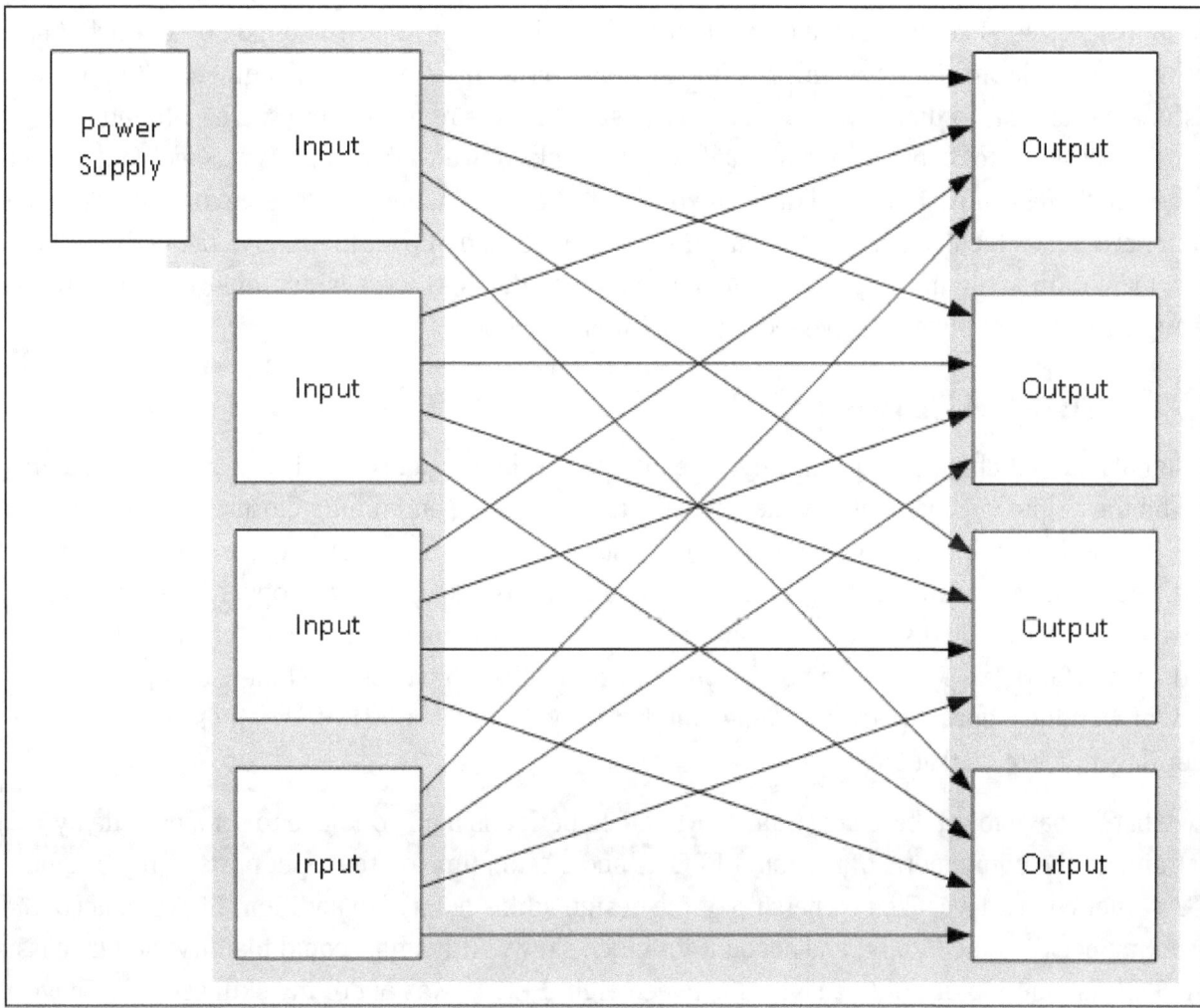

Figure 48: PCB Layout of FPGA with Power Plane

Fixing this required some creative thinking. Luckily I had a very skilled and experienced bench technician. We had just moved into a newly retrofitted building. Why is that important? The electricians had left behind some of their "trash". In the trash pile was some 12 gauge wire. The same stuff you have in the walls of your house. Being a skilled and experienced bench technician, he squirreled some of this away. After some analysis and bouncing around a couple of ideas. He stripped down some of the wire, and with me looking up decoupling capacitors on the schematic he soldered the 12 gauge wire on the back of the board making connections to each FPGA. Adding the wire to the power signal significantly reduced the resistance of the power distribution and eliminated the voltage drop between the power supply and the FPGAs. Problem solved, or so I thought.

With all FPGAs loaded and adequately powered, I could put a signal into each of the four inputs and see an output signal on all four output ports. It was not right, though, and it changed over time. This turned out to be a few problems that I started working on simultaneously. When things change over time, I always suspect a thermal issue. I suspected a thermal issue when I saw high currents on the power supply. After all, whatever power goes in has to be dissipated. I confirmed this by sticking my hand in the chassis. I could feel the heat radiating from the parts. I know enough not to touch them. Complex power dissipation problems require consultation with a mechanical engineer with heat transfer experience. Fortunately, this was a pretty straightforward problem to solve. I found a manufacturer with stock for a heat sink made for the package we were using. It was well documented, and the parameters showed it could dissipate the heat. I ordered them and gave them to my skilled and experienced bench technician, who had taken some time to study up on thermally conductive adhesive for heat sinks, and with the help of a fan in the chassis, the FPGAs were within their temperature range. Although the heat problem disappeared, I knew I was not done.

3.2 Timing Analysis

The signals did not change over time anymore but they still were not right. There was still random noise on the output signals. For me, the next suspect is clocking and timing closure. As you have seen so far, setting constraints and achieving timing closure is part of the design process. My thought was that the constraints that were set were not tight enough or just plain wrong. I opened the constraint file and was again staring in disbelief. The only constraint that was set was the clock constraint which, luckily, was 200MHz. There were no IO constraints on either chip design. None! This produced plenty of warnings, during place and route, but there was a large number of warnings and they were camouflaged, more on that later.

Now what? I have to set these constraints. My clock period is 5 nS so I have to set a prop delay constraint on the output ports of my input FPGAs and a setup time on the input ports of my output FPGAs (notice that I've taken ownership of the design at this point). In addition, I have to account for the routing between the FPGAs and account for clock skew. Although I could identify the plane issue in the PCB layout, looking at 512 signals between eight FPGAs was out of my skill set. I did have a skilled and experienced PCB designer on speed dial. As an aside, the PCB designer and the bench technician were good friends and ended up starting their own company shopping their services around to small startups. The PCB designer was able to read the design files and, with some finagling, produced a net list of the signals between the FPGAs with associated lengths. I was able to convert the lengths into time delays. More bad news. In a system like this, it is desirable to make the delays between the banks of FPGAs as consistent as possible. In this case, the trace lengths had significant differences. Some of the longest delays were more than twice the length of some of the shortest. There was also a lot of clock skew.

I started a big spreadsheet like the one in Table 1. With some experimentation, I found a safe minimum for the clock to Q constraint on the output ports of the input FPGAs and the setup time of the input ports on the output FPGAs. I entered that into the spreadsheet along with the wire delays and the clock skews. The clock routes on the PCB were not balanced either. A total of 512 signals with 8 different clock skews. It took a while to get this right, but when I was done, there were some negative slack numbers.

In parallel, I took what I had estimated as a safe clock to Q and setup delays and entered them as constraints in the respective FPGAs. Unfortunately, I was having trouble with timing closure. The culprit was the internal clock skew. I checked, and this should not be the case because there was a clock manager that should have been deskewing the signal. Remember when I said it is hard for an experienced designer to find rookie mistakes? This time it was hard for me to find the problem. The designer had connected the feedback input of the clock manager directly to its output, not to the output of the clock buffer. The result was that the clock manager was ineffective at removing the clock skew. It also produced a warning that was ignored. I fixed this. It turned out to be a single character change in the code. This change had the most profound effect on my journey to get this design working. At room temperature, the design was working correctly.

There was still some negative slack. I knew that this would likely cause problems with temperature and the testing I had done was with the chassis uncovered. I was afraid that once the cover was on and the temperature rose a little or it was in a hotter than normal room those negative slack values would be a problem. There is an option in the clock manager to add a skew between the incoming clock and the internal clock. I played around with this and found a case where the skew values were positive. This was an empirical exercise, and I do not recommend doing this. At one point, I thought I would have multiple versions of each design, each with a different skew. I managed to avoid that.

Before continuing, I should address the general issue of warnings when using these tools. Any error that is produced will stop the process. Warnings, on the other hand, will allow the design to continue but should be addressed by the designer. This design had a lot of warnings that were benign, like unused signals. All of these warnings could have been eliminated so the important warnings, like missing constraints or an improperly connected clock manager, are more obvious in the list. Also, any warning that warns that there is a possible simulation/synthesis mismatch should be taken very seriously because it may be a case where the verified design is different from what is being built.

One last thing. The boards did not always respond properly to power-on reset. As expected, there was no synchronization on the incoming power-on reset. I should have known.

3.3 Final Result

Although it probably only took about fifteen minutes to read my account, it took me about four weeks to complete the work. This was enjoyable work, and I learned a lot. It is hard to say it was completely enjoyable because we were under intense schedule pressure when I started, so there were some late nights during the four weeks. There was some management pressure as well. The final working prototype was done just in time to demonstrate it to a few critical customers and win some sales.

To make this a producible product, there was no question that the PCB needed to be redesigned. My skilled and experienced PCB designer made the voltage plane continuous with a more than adequate connection to the power supply. In a reasonable number of layers, he could route the 512 signals

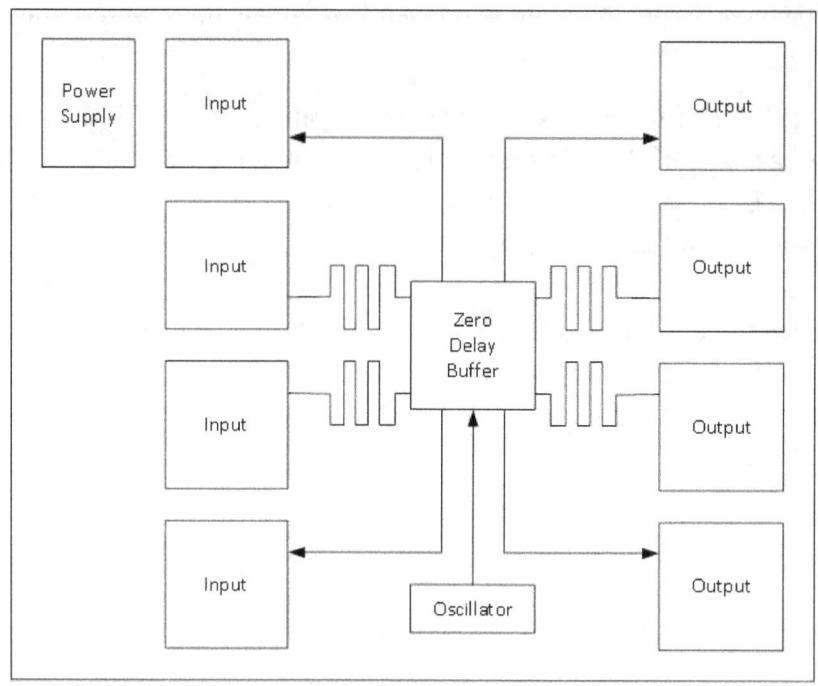

Figure 49: Clock Distribution with a Zero Delay Clock Buffer

between the FPGA with a very small length difference. There was less than 100 pS of skew between the longest and the shortest. This allowed me to remove that extra skew and have a positive slack on all signals. I also added a zero delay clock buffer as shown in Figure 49. The buffer was located as close to the center of the array as possible. The routing from the buffer to the FPGA was balanced, so all of the clock traces were the same length which required some of the traces to have a serpentine route. The result was that all of the FPGAs saw the active edge of the clock at the same time. The board worked on the first try and was able to ship basically on time. As for the consultant, he got paid most of his fee. Notice that the only changes I made to his code were to fix the feedback on the clock manager and to add the reset synchronizer. His DSP algorithm worked flawlessly and was the heart of the product. A couple of years later, my phone rang. It was the consultant. I hesitated but answered. At this point, I was on to my next startup. After some pleasantries, he asked me if I remembered the system that I have been discussing in this chapter. Of course, I did. He remembered that I had made a change that made a significant difference in making the system work right. I offered to meet him at his office the following Saturday for an agreed-upon fee. I showed up, looked through the code, and found the line where the feedback of the clock manager was connected to the output of the clock manager, not the output of the clock distribution network. I fixed that. For good measure, I also put in a reset synchronizer. He recompiled and loaded it into the system, and it worked perfectly. He was ecstatic. I am pretty sure he was under the gun to get this out. At this point, 30 minutes had passed out of my promised 8 hours. His son, who had just graduated with his EE degree, was there. I spent the rest of the day going over good design practice and running through the slides of my FPGA class with him. I keep in touch with both of them.

3.4 Exercise

1) It has been said that "Those who do not learn history are doomed to repeat it[1]". Write a case study of a project you completed. Include setbacks, problems solved and personal interactions that were part of the process of completing the task.

---

[1]   This quote is derived from the quote, "Those who cannot remember the past are condemned to repeat it", by George Santayana.

4. Physical Design and Simulation

Physical design is based on the gate-level abstraction of the FPGA. The gate-level abstraction is the synthesizer's output and is not formatted to be human-readable. The gate-level abstraction is a netlist file made from components available on the FPGA. It is used as an input to the simulator. More importantly, the gate-level netlist is the input to the place and route (implementation) tool. In addition to the place and route step, the implementation tool verifies timing closure using static timing analysis (STA). A Standard Delay File (SDF) is an output of the implementation, which is used for gate-level simulation. The diagram in Figure 50 is a generic design flow for FPGA development. The left side is the physical design process, and the right side is the simulation process which is not covered in depth in this text. The combination of synthesis and implementation is commonly referred to as compilation.

As noted earlier in Chapter 0, the process starts with a code base. The progression on the left side starts with synthesis, which is covered in Section 4.1 followed by Place and Route (Implementation), which is covered in Section 4.2. Both of these processes are informed by Spatial and Temporal Constraints. The generation and use of the programming file are covered in Chapter 5

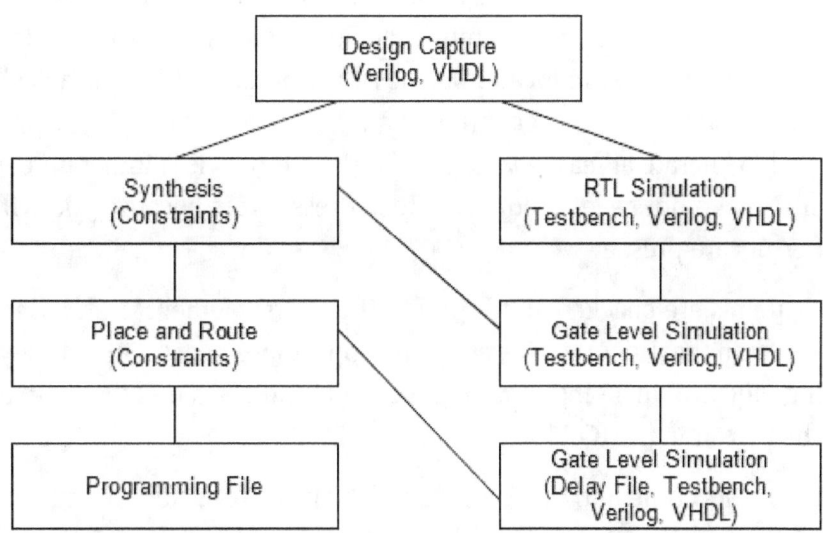

Figure 50: Generic Design Flow

4.1 Synthesis

Synthesis is the conversion of a Register Transfer Language (RTL) model into a gate-level model. A designer should always visualize what will be implied by the code they are writing. This is where the adoption of coding guidelines pays off. Although the RTL is the obvious input to the synthesizer, the constraint file is also essential. The synthesizer needs to know what components it should translate the RTL into. For that reason, the manufacturer/family constraint is essential. Also, the synthesizer needs to know the quantity of resources available and it needs the die designation for that. The package constraint is also necessary because the synthesizer needs to know how many pins are available. The one Temporal Constraint necessary in synthesis is the clock period constraint. It will give a general limit on the amount of logic that is synthesized between Flip-Flops. If the synthesizer predicts that the design will not make timing, it is almost guaranteed that the implementation step will show timing violations. If the synthesizer predicts the design will make timing, the implementation step will be run, and static timing analysis is run to verify that there are no timing violations.

4.2 Place and Route (Implementation)

If the code is carefully written for synthesis, the synthesis process is pretty routine. The implementation step relies heavily on all of the constraints. Once constraints are set, it may be possible to run implementation to completion without any other input. In designs with unique characteristics like low power, small size, or high performance, strategies for implementation are selected and modified.

A baseline for the implementation step is the pin location constraints. The pin locations force the placement of signals and will drive the overall placement of the components in the gate-level netlist. If a pin location constraint is missing, the implementation tool will assign it to an available pin. It is important to note that this is not the most optimal location for the board design and most importantly if the design changes and synthesis is rerun, the implementation step may place that unconstrained pin in a different location. A signal whose pin location is moving around could cause a design error that would be difficult to find and may be unrecoverable on the PCB design. If this happens, the implementation tool will generate at least a warning, if not an error. It is imperative to address all warnings in a design. You do not have a choice but to address errors because you will not get a completed design if errors are present.

The input-output constraints are checked at this point. If the first element in an input chain is a Flip-Flop and the last element in the output chains are Flip-Flops, these constraints, if they are feasible, should pass. A timing violation in an input or output usually indicates that the logic design did not allow a Flip-Flip to be placed in the IO block.

Finally, the clock constraints come into play. These are the basis for static timing analysis of the implemented design. It is possible that if the techniques described in this book are followed carefully, this step will complete without errors or warnings. Timing violations in the implemented design are addressed using the techniques described in Section 1.9. If those techniques are attempted, along with anything creative you can think of, there is still an option.

Most place and route tools allow you to select a strategy to optimize the result for size, power, or performance. Usually, the default strategy of the implementation tool will be best. It is designed to balance size, power, and performance. If there is a desire to fit in a smaller die size in the selected package, optimizing for size may get the design within that die. Size optimization comes at the expense of performance. Also, if you are optimizing for size and just fit within a small die, a design change may make it necessary to go into a larger die which could mean removing a part from the board and installing a larger die in the package. Worse yet, if you are trying to squeeze into the largest die for your selected package, there is a risk that you will not be able to fit into the package at all.

Another axis of optimization is the performance which boils down to a faster F_{MAX}. Using this optimization strategy to speed up F_{MAX} will most likely increase the number of gates utilized in the FPGA. The thought is that once you have bought the FPGA you may as well use all of the gates. The downside, however, is power. A strategy that optimizes for higher performance and increases utilization inevitably increases power consumption. Conversely, a design optimized for lower power will have lower utilization and lower performance. The selected strategy, if not the default, will be stored in the constraint file along with the other physical constraints.

Engineering is all about trade-offs. Understanding the trade space, which in our case is defined by constraints, will give you the best chance of producing a cost, power, and space-efficient FPGA design.

4.3 Physical Simulation

Although verification is not treated in this text, it is worth discussing the simulation process. When the RTL abstraction is simulated, the model will logically function the way the code was intended. There will be no actual time delays, though. All of the delays will be the same, small unit of time. Figure 51 shows a case where a byte is clocked out of a series of Flip-Flops. The Flip-Flops use the rising edge as the active edge. In this example, the byte in the Flip-Flops before the active edge of the clock is 0x38. After the rising edge, the value at the output of the Flip-Flops is 0xC7. The important thing to note is that all of the bits change a single time unit after the clock.

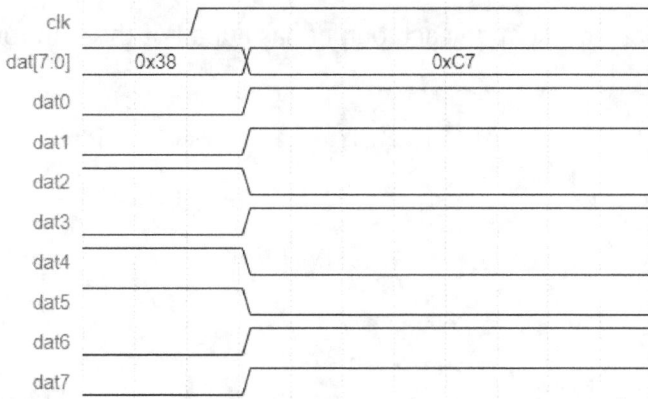

Figure 51: Unit Time Simulation Results

In a real-time gate-level simulation, there is an estimation of the delay of each bit that comes from the place and route process. This is the Standard Delay File produced by the Implementation step. If timing closure has been achieved, the function of the real-time gate-level simulation is the same as the unit-time simulation. Figure 52 is the same function as the example shown in Figure 51. The difference is that the time delay after the clock is longer than the unit-time example and differs from bit to bit. This shows as a period of uncertainty. Looking at the separate bits, the differences in the delays are obvious. The simulation duration for a real-time simulation is longer because the simulator has to perform calculations with more detailed elements and also has to handle more time calculations because the delays are different.

Figure 52: Real-time Simulation Results

4.4 Exercises

1) Download and install the design software for the component selected in Exercise 2.1.

2) Modify the code in Figure 53 by changing adder_m to maximize the number of IO that are used in the component selected in Exercise 2.1 and change adder_n to maximize the amount of internal resources that are used. This will take some trial and error.

3) What is the estimated F_{MAX} of the solution for Exercise 4.2?

4) What is the estimated power assuming each output has a load of the solution for Exercise 4.2?

```
library IEEE;
use IEEE.STD_LOGIC_1164.all;
use IEEE.STD_LOGIC_ARITH.all;
use IEEE.STD_LOGIC_UNSIGNED.all;
entity adder is
  generic (adder_m : integer := 16);
  port (adder_clock        : in  std_logic;  -- System clock input
        adder_reset        : in  std_logic;  -- System reset input
        adder_input_busa  : in  std_logic_vector(adder_m - 1 downto 0);  -- Input Bus A
        adder_input_busb  : in  std_logic_vector(adder_m - 1 downto 0);  -- Input Bus B
        adder_output_busa : out std_logic_vector(adder_m - 1 downto 0);  -- Output Bus A
        adder_output_busb : out std_logic_vector(adder_m - 1 downto 0)   -- Output Bus B
        );
end adder;
architecture rtl of adder is
  constant adder_n   : integer := 45;
  signal adder_i     : integer;
  type reg_array is array (1 to adder_n) of std_logic_vector(adder_m - 1 downto 0);
  signal adder_tempa : reg_array;
  signal adder_tempb : reg_array;
begin
  adder_Process : process (adder_clock, adder_reset)
  begin
    if (adder_reset = '1') then
      gen_rst : for adder_i in 1 to adder_n loop
        adder_tempa(adder_i) <= (others => '0');
        adder_tempb(adder_i) <= (others => '0');
      end loop;
      adder_output_busa <= (others => '0');
      adder_output_busb <= (others => '0');
    elsif (adder_clock'event and adder_clock = '1') then
      adder_tempa(1)      <= adder_input_busa + adder_input_busb;
      adder_tempb(1)      <= adder_input_busa - adder_input_busb;
      gen_reg : for adder_i in 2 to adder_n loop
        adder_tempa(adder_i) <= adder_tempa(adder_i -1) + adder_tempb(adder_i -1);
        adder_tempb(adder_i) <= adder_tempa(adder_i -1) - adder_tempb(adder_i -1);
      end loop;
      adder_output_busa <= adder_tempa(adder_n) + adder_tempb(adder_n);
      adder_output_busb <= adder_tempa(adder_n) - adder_tempb(adder_n);
    end if;
  end process adder_Process;
end rtl;
```

-(Unix)--- **adder.vhd**   Bot L41   (VHDL)

Figure 53: Exercise 4.2 Listing

5. Hardware Design

Hardware design comprises the decisions that must be made at the board level. I am leaving out printed circuit board construction and assembly techniques because they are well covered in other textbooks written by mechanical engineers and material scientists. Instead, I am focusing on FPGA-specific aspects centered around transitioning the design abstraction to the actual hardware and getting the design working on the real hardware the way it was predicted to work in simulation.

5.1 Configuration and Programming

Let's get a couple of definitions out of the way. Configuration is the method the FPGA uses to hold the logic design. Programming is the method the system uses to get the synthesized and placed and routed HDL design into the FPGA. The final output of the synthesis and place and route steps is a file containing a data stream that defines how each programmable element in the FPGA should be set. This file varies based on the programming method and manufacturer.

Modern FPGAs hold the design using one of two configuration methods, EEPROM and SRAM[2]. EEPROM devices are non-volatile, which means when they are powered down and back up they are still programmed. They are also lower power because EEPROMs draw very little power to hold their state. This makes sense because they can be powered off and not lose their program. The downside of EEPROM-based FPGAs is density. EEPROM cells are relatively large, so the number of logic cells per unit area on the FPGA is less than it could be. It is much more common for FPGAs to use Static Random Access Memory (SRAM) to solve the density problem. SRAM cells are much smaller than EEPROM cells, so SRAM-based FPGA have more cells per unit area. SRAM based FPGA are also faster because the logic cells are closer together. This comes at a cost. SRAM cells draw a little more power than EEPROM cells. The more significant downside is that SRAM cells are volatile. Every time an SRAM-based FPGA is powered up, it must be programmed. There are trade-offs to be made in programming methods to mitigate this.

There are a few options for programming, and a given design may accommodate more than one. The most common for SRAM-based FPGAs is to put a Serial EEPROM (SEEP) on the PCB and connect it to the FPGA. The connectivity of this is shown in Figure 54. Sometimes the EEPROM is integrated into an FPGA. On power-up, the FPGA will load its configuration from the SEEP, check that there were no errors and begin operation. The SEEP, of course, is non-volatile. This might seem counter-intuitive. Why would you use an EEPROM to configure an SRAM-based FPGA instead of using an EEPROM FPGA? The answer is based on the fact that a Serial EEPROM is an array of EEPROM cells. This arrangement is much more efficient than distributing EEPROM cells throughout an FPGA. Using an SRAM-based FPGA with an external SEEP provides a much denser FPGA with the addition of a small and cheap external SEEP that employs a relatively simple interface for FPGA programming.

---

[2]   There are a small number of FPGAs that are One-Time Programmable (OTP). They are used mainly in applications where other (configuration) methods cannot be used, including space applications or high radiation environments.

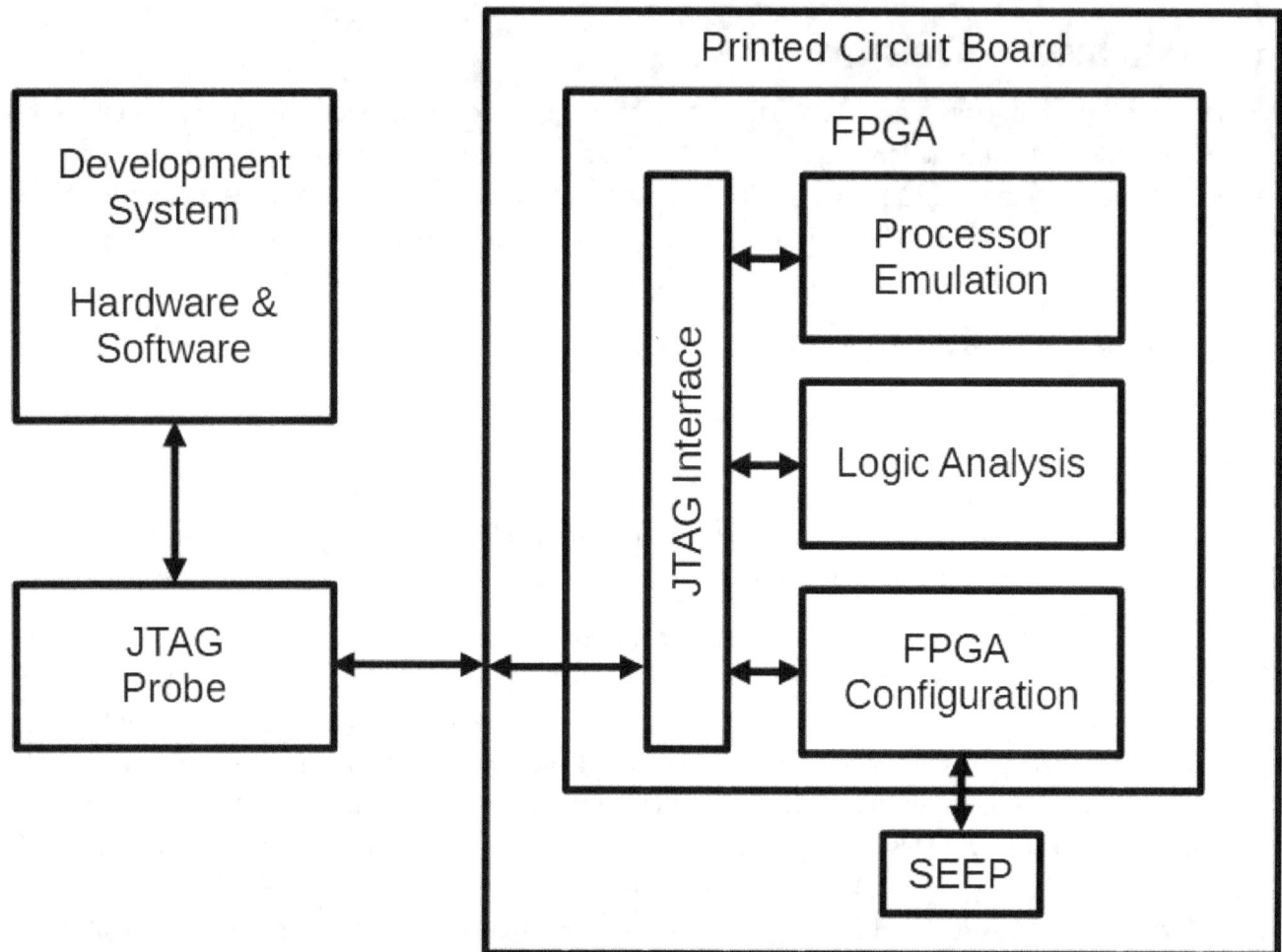

Figure 54: JTAG Support for Development Tasks

Whether it is an EEPROM-based FPGA or an SRAM-based FPGA with an External SEEP, the non-volatile part needs to be programmed. There are two ways of accomplishing this. A standard method is to use a programming cable connected to the design system. These cables usually connect to the computer running the design software using a USB port. There is a small amount of interface circuitry in the cable, and the other end connects to the FPGA using a manufacturer-specific connector. It is commonly the case that the small amount of interface circuitry is put on the PCB itself and the connection to the computer is simply a USB to USB cable. This is simpler to use, but in very small systems or very cost-sensitive systems, the addition of the interface circuitry may be prohibitive. Once connected, the design software transfers data to the FPGA or to the SEEP through the FPGA. The exact details of these methods vary from manufacturer to manufacturer and even between families within a given manufacturer's offerings. Consider the details of the methods and the specific options carefully and work closely with the PCB designer to implement them. Also, make use of any reference information that the manufacturer provides. A less common method of programming the SEEP is to program it before it is installed on the board. The part distributor usually does this. To achieve this, the customer specifies the bit stream, and the distributor programs the SEEPs before delivering them. Preprogramming is only advantageous in a high-volume production environment.

The programming cable mentioned above can directly program an SRAM-based FPGA. This bypasses the need to program the SEEP. During the bring-up and debugging of the system, see Section 5.2, there may be the need to change the logic design frequently. Programming the SEEP can take minutes, whereas programming the FPGA directly from the programming cable is done in seconds. This is only done in the lab during bring-up and debugging. It should be noted that although writing a SEEP is relatively slow, reading it is fast. It is faster than using the programming cable. The actual time to transfer the data from the SEEP into the SRAM-based FPGA should be accounted for in the time it takes the system to initialize. Also, there is almost always an error-checking step that the FPGA performs on the data. There may be an option to compress this data to save memory space which can also take time during initialization. Some FPGAs also provide an option for data encryption to secure the logic design, which can add additional time overhead.

The final programming method requires an onboard microprocessor system. It is common for a digital system have a large FPGA and a separate microprocessor. In this case, a part of the microprocessor's memory is allocated for the FPGA image. There are two specific advantages. First, there is only one memory chip, so it costs less and takes up less board space. Secondly, the processor's memory is usually externally upgradeable. It is common for the software in a microprocessor system to change for bug fixes or feature enhancements. If the FPGA logic design is also in that memory, it is upgraded when the software is upgraded. The disadvantages can be significant. One system aspect that must be worked out is that the operation of the microprocessor cannot rely on the FPGA being programmed. Also, and more importantly, the FPGA data stream needs a lot of memory. It will likely impinge on the software image's ability to grow. Planning for this early in the system design process can alleviate problems that could come up at this stage.

5.2 Bring-up

Bring-up comes in two parts, the planning you do during PCB design and the actions you take in the lab when the PCB arrives. There are three main things that are not modeled by the implementation tools or checked in the simulation. They are power, clock, and reset. These things should be dealt with in that order. A high-level schematic of a notional power supply, a voltage monitor, and an oscillator is shown in Figure 55.

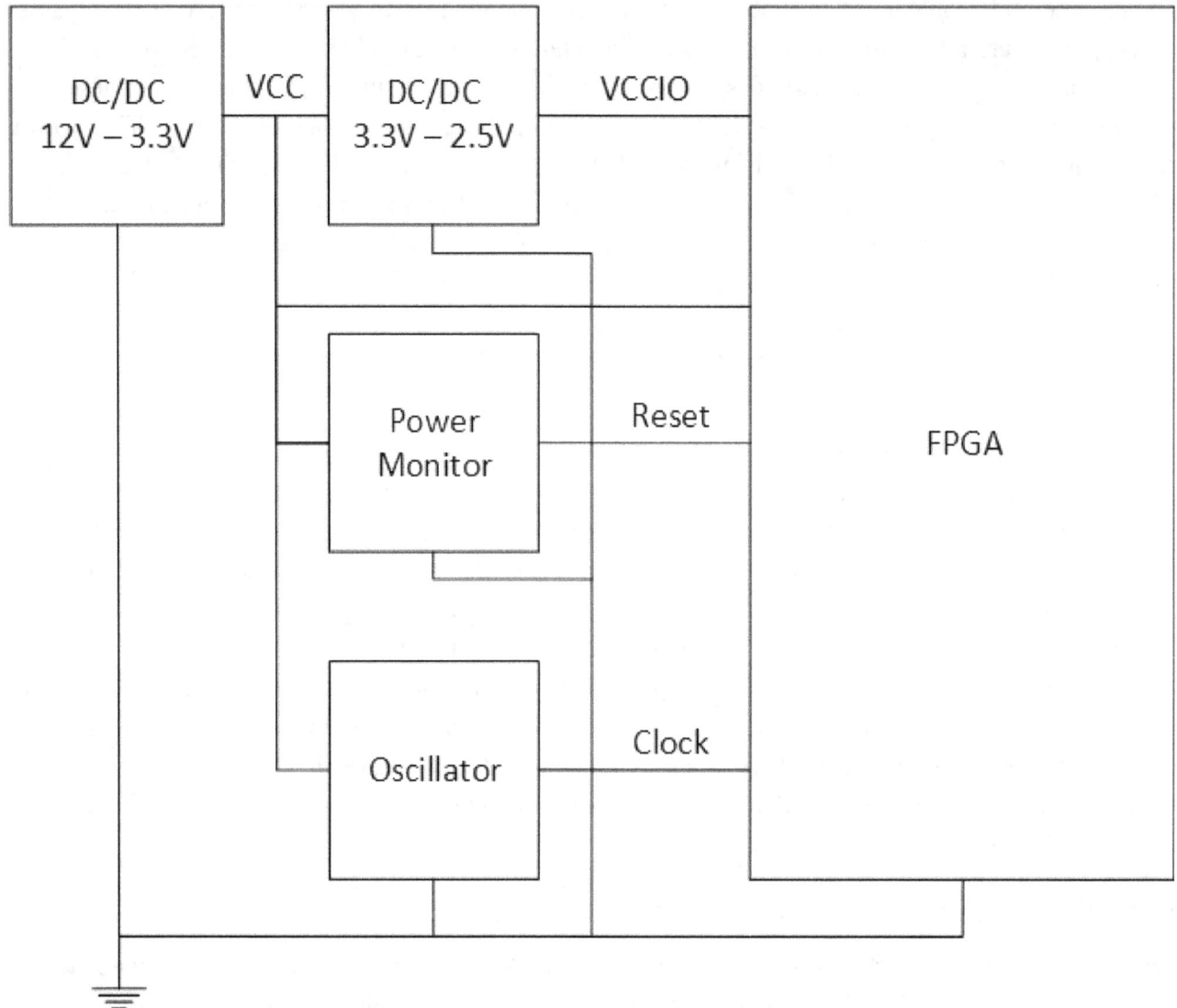

Figure 55: Notional Power, Clock, and Reset

FPGAs usually have more than one power supply, one to power the core, one for each IO bank, and sometimes power supplies for special components like clock management devices. It is critical to adhere to the manufacturer's requirements for power supply regulation, stability, and decoupling. The power requirements of these supplies are not as predictable as they would be in an off-to-shelf component like a microprocessor or a memory. The actual power draw of an FPGA is based on the design you put in it. For this reason, you must size the power supply based on calculations done in the implementation tool, plus some margin considering enhancements to the design. It is not out of the question that if you utilized 20% more of the resources in a given FPGA, the power draw could go up by 20%. I am not suggesting that you size your power supply for 100% utilization of the FPGA. I am recommending that you leave some margin for future growth. I have seen at least one design where the FPGA was designed, and less than 50% of the component was utilized. The power supply was sized accordingly. At some later time, the system requirements changed, and significantly more of the FPGA was utilized. The designers rejoiced that the new function was successfully implemented in the existing FPGA. Reality sunk in when the FPGA was programmed, and the power supply could not maintain regulation.

Additionally, and maybe most importantly, the time delay from turn-on to a stable power supply is significant to know and plan for. Power converters can have turn-on times from tens of microseconds to a millisecond. It depends on the topology of the converter and how much power it is expected to supply.

When there are multiple power supplies, the FPGA manufacturer may specify the sequence that they must be brought online. Datasheets might be subtle in the way they specify the sequencing requirement in that they do not usually just come out and say a given supply must be on before another is turned on. They will generally show a voltage difference specification requiring that the voltage on a given supply never exceeds the voltage on other supplies. These are ramps, not step functions, and the ramp may not be linear.

Bring-up is also a function of clocks being stable. Digital designs have at least one external clock, which is the time base for the synchronous logic in the design. Usually, the actual clock or clocks are derived from this clock using clock management devices on the FPGA. Oscillators used to generate these clocks have a turn-on time as well. The measurement of the turn-on time starts when the power supply is in regulation. Two parameters must be met for an oscillator to be considered stable enough. The first parameter is the amplitude. Oscillators start generating an oscillating output very soon after power is applied. The amplitude is very low initially but eventually ramps up to its full value. Secondly, the oscillator's frequency is generally not within the specified range right after power up either. Very high-precision oscillators can have very long turn-on times. The general-purpose oscillators commonly used in FPGA-based systems have turn-on times on the order of tens of milliseconds.

When the system is waiting for power and the clocks to meet their specified requirements, the FPGA is held in reset. All FPGAs should have a single reset input that puts all Flip-Flops into a known (usually logic 0) state. Reset on the PCB is generally managed with a voltage sensor that senses that the power supplies are in their expected range and starts a timer. When the timer has elapsed, the voltage sensor removes the reset signal and allows the FPGA to begin running. If the power supply goes out of regulation before the timer has elapsed, the timer restarts. Note that this reset signal is not synchronized to the clock. This must be handled by a reset synchronizer in the actual FPGA design as described in Section 1.6 on Page 27. A timing diagram for power, clock, and reset is shown in Figure 56.

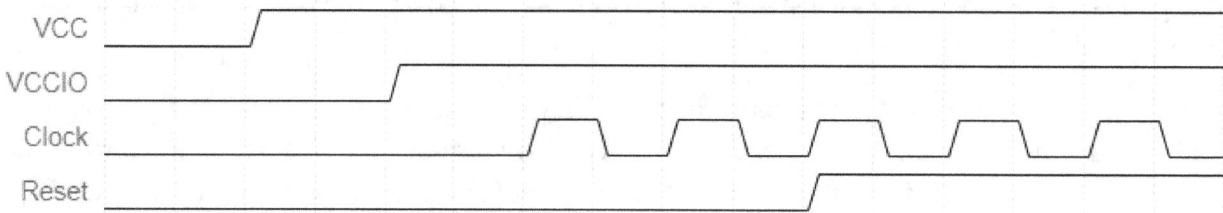

Figure 56: Power, Clock, and Reset Timing

A table showing approximate times for power supply start-up, oscillator start-up, and reset delay is shown in Table 4. Note that these times are approximate and will be based on the specific component selection. The reset time defines the total time because it starts when the power supply is in regulation and ends at a safe time after the oscillator is started.

Table 4: Component Startup Times

Component	Startup time
Power Supply	0.1 – 1 mS
Oscillator	10 – 100 mS
Reset	100 – 500 mS
Total	101 – 501 mS

I cannot remember how many times if heard a designer say, "Well it worked in the simulation". My answer is always that simulation does not cover power, clock, or reset. I think from now on, I will just keep a copy of this text handy and give it to them when they cannot understand why a design that worked in simulation does not work reliably in the lab.

The design is done, the board is built, and it is in the lab, now what? Turn on the power supply, see the system flawlessly perform all of its functions, and go out to celebrate! That would be nice. The reality is that if you followed all the steps required for first-pass success, the system might just work. Be careful to define what "working" means, for example, temperature range, intermittent behavior. Considering there could be a problem, and there usually is, you should follow a measured, sequenced approach to bringing up the system in the lab.

Most importantly, have a plan. It does not need to be a meticulously formatted and reviewed document. It may just be a checklist. In a situation where excitement, frustration, and fatigue may set in, it is good to have a reference for the next step. Here are some steps to follow.

- Assemble the equipment you need and reserve the necessary lab space well before you expect the assembled printed circuit boards to be delivered.

- Inspect the assembled printed circuit board before you apply power. Small things can creep in during the first assembly run. Ensure that non-populated components are not populated.

- Perform simple electrical checks like checking to see if power is shorted to ground. Note that the resistance between power and ground may be minimal and appear to be shorted.

- Have a peer assist you. Another set of hands, another set of eyes, and another experienced brain can help in the first hours of bring-up.

- The PCB designer should be on speed dial if not sitting next to you.

- Always check the actual voltage of a lab power supply with a multi-meter before connecting it to your lab set-up.

- I try to take pictures of the lab setup before I start in case there is a question about how things are connected or what equipment is used.

- When power is first applied, monitor the state of at least one of the supplies going to the FPGA. If it is not right, turn the power off (it should now be obvious why the second set of hands is helpful).

- Have the design system available on the lab bench. Try to program the part once you are confident the power supplies are in regulation and the clocks are running.. If it does not work on the first try, the design tool may provide some helpful information on why it did not.

- Your list will vary based on your specific application, design implementation, and the lab equipment you have access to.

5.3 Debug

Once the system is loaded and generally performing its intended functions, you will likely find bugs. The best case is when you can isolate the bug, adjust your testbench to repeat it, find the problem, solve it, and test the solution. The simulation environment has the ultimate controllability and observability. If it is not practical to replicate the bug in simulation, usually because it is too time-consuming, you will need to debug it in the lab.

FPGAs present a few options for controllability and observability not available in other components. FPGAs have observability options using internal logic analyzers. This component is compiled into the chip and connected to internal nodes. An example showing an internal logic analyzer installed in an FPGA is shown in Figure 54 on page 70. Unused internal memory is used for the logic analyzer capture buffer. The Graphical User Interface (GUI) is part of the design system. Finally, the cable used for programming also acts as the connection between the design system on the PC and the board. These logic analyzers have essential functions like those available on conventional logic analyzers, including triggers and display options. Because the analyzers use memory that is otherwise unused, these analyzers are limited in the number of signals they can trace and the depth of the traces. That is considered a minor limitation when you consider that a full-featured logic analyzer would require a large number of difficult to make connections to probe the signals. In a very high pin count, a dense package that is common within FPGA components, this could be extremely difficult and could end up being impossible. It should also be noted that it is possible to create very complicated triggers by writing triggers in HDL and including them in your code as a trigger connection to the internal logic analyzer. For controllability, some FPGA families and their associated design systems can make low-speed input connections inside the FPGA that are also controlled by the design system.

A quick aside. The interface used for programming and now the controllability and observability for debugging can also be used for processor debugging if there is a processor embedded in the FPGA. This interface is an extension of chip-level emulation and diagnostics that started to be included on an embedded processor when their packages got too large to accommodate board-level emulation and diagnostics.

It's not a bad idea to set up the logic analyzer before you bring the FPGA to the lab for the first time. It will get through any glitches in the process. Significant top-level control signals are good candidates for the first set of signals to look at. Also, even if the design appears to work, it is helpful to look inside just to ensure things are working as expected.

Figure 54 on page 70 shows the various development tasks the development system supports through JTAG. It shows a development system made up of a personal workstation (Windows or Linux) running the manufacturer's development software connected to a JTAG probe a manufacturer or a third party supplies. This device converts an interface from the development system, usually USB, to the JTAG interface. Sometimes this interface is on the PCB. This combination supports the following tasks:

- FPGA Configuration is the data stream transfer from the place and route tool to program the FPGA.

- Logic analysis of the FPGA to improve the controllability and observability of the FPGA.

- Processor emulation to load software and to control and observe the operation of a processor contained within the FPGA. This can be a soft processor synthesized into the FPGA or a hard processor that is part of the FPGA die.

A nice addition to the board design are test points that can be used to connect test equipment during bring-up. If possible, a test point on each power supply, each clock, and the reset signal should be included. Ground test points associated with each point should also be included as much as possible. If there are two to four unused pins, those could each be connected to a test point with an associated ground pin. These allow for a little controllability and or observability without configuring an internal logic analyzer.

5.4 Exercises

1) Identify the clock, reset, input and output signals of the development board selected in Exercise 2.1.

2) Compile either the SystemVerilog or VHDL code shown in Figure 29 targeting the development board selected in Exercise 2.1. Be sure to set the appropriate temporal and spacial constraints. Verify that all constraints are met.

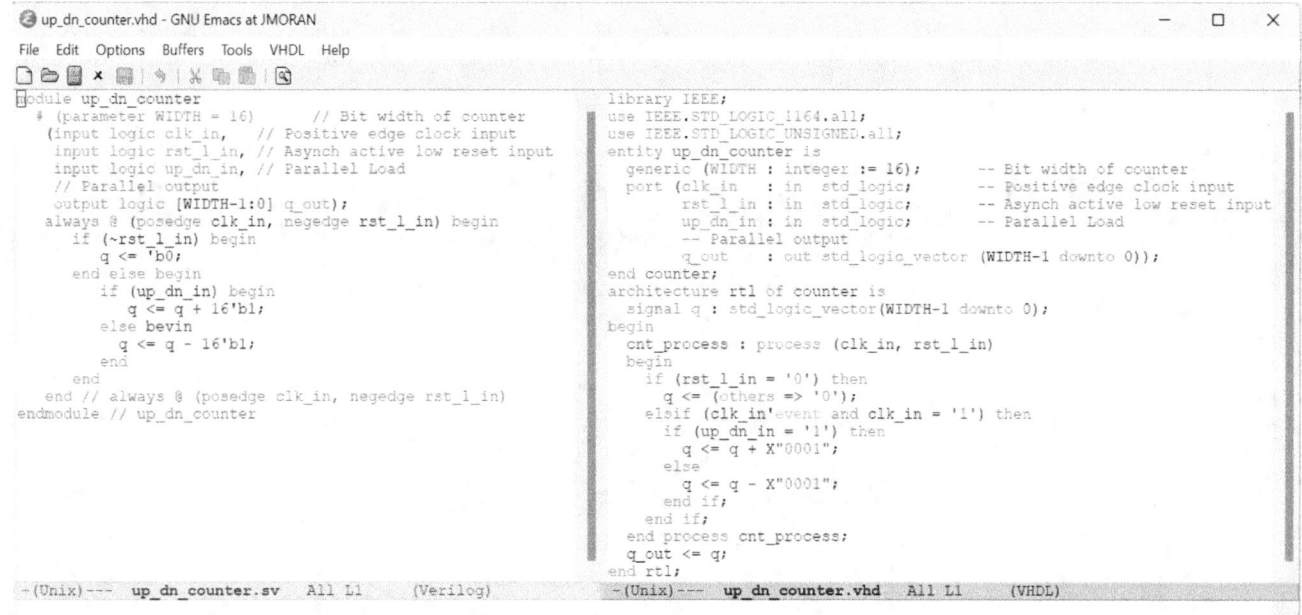

Figure 57: Up/Down Counter Listings

3) Instantiate an embedded logic analyzer and recompile the design created in Exercise 5.2.

4) Load bitstream generated in Exercise 5.3 into the development board and verify its operation using the embedded logic analyzer.

6.   Tool Chain Staging

There are several tools involved in FPGA design. Chapter 4 describes the software involved in converting the HDL design into a bitstream that is loaded onto a board. Chapter 5 describes how to put that bitstream onto a board, manipulate it, and gather information for debugging. Major FPGA manufacturers support these functions from a single integrated development system. Because of the number of functions the tool must support, it ends up being more than a single mouse click to go from HDL to a working FPGA design. Learning these tools requires taking classes from the manufacturer, watching third-party videos, and, most importantly, using the tools.

Developing proficiency in the tool is difficult when it is coupled with the complexity of developing and implementing a design. Staging the tool by running through the process with a familiar, simple design is the best way to try out the features. An example of this is the counter shown in Figure 58. This particular counter has a clock and reset, three control lines, sixteen data inputs, sixteen data outputs, and a single status output. Running this code through the process is enough to stage all parts of the FPGA design process.

6.1   Synthesis and Implementation Staging

The first step in developing an FPGA is to select a manufacturer/family/device/package combination. To stage the tool correctly, the same manufacturer/family/device/package combination used in the target design should be selected for staging. Component selection is the first constraint. The next set of constraints must be constraining pin connections for each input and output in the entity. If this FPGA is targeted to a reference board, connect the clock to the onboard oscillator and the reset signal to a push button if available. Remember that the reset must be synchronized to the clock described in Section 1.6. This is also an excellent time to set T_{SU} and T_H constraints for the inputs and T_P constraints for the outputs. Add a frequency constraint on the clock to round out the constraint file.

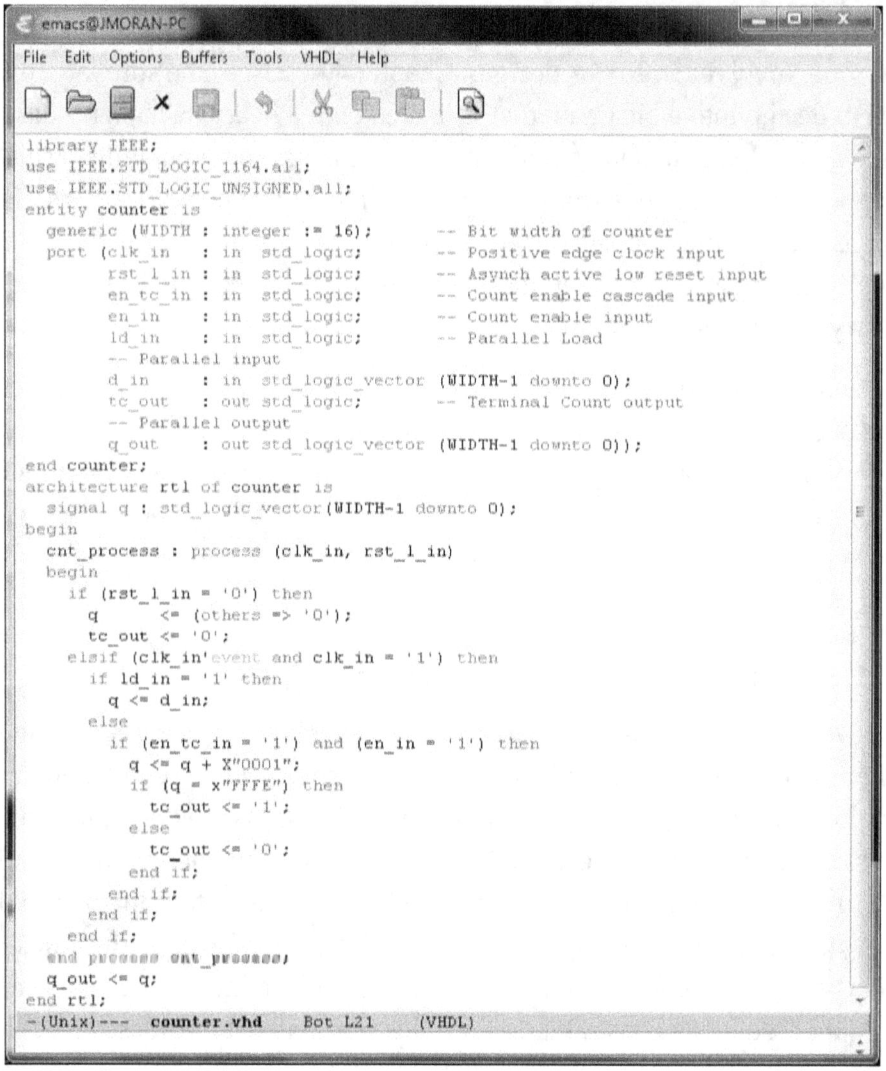

```vhdl
library IEEE;
use IEEE.STD_LOGIC_1164.all;
use IEEE.STD_LOGIC_UNSIGNED.all;
entity counter is
  generic (WIDTH : integer := 16);          -- Bit width of counter
  port (clk_in    : in  std_logic;          -- Positive edge clock input
        rst_l_in  : in  std_logic;          -- Asynch active low reset input
        en_tc_in  : in  std_logic;          -- Count enable cascade input
        en_in     : in  std_logic;          -- Count enable input
        ld_in     : in  std_logic;          -- Parallel Load
        -- Parallel input
        d_in      : in  std_logic_vector (WIDTH-1 downto 0);
        tc_out    : out std_logic;          -- Terminal Count output
        -- Parallel output
        q_out     : out std_logic_vector (WIDTH-1 downto 0));
end counter;
architecture rtl of counter is
  signal q : std_logic_vector(WIDTH-1 downto 0);
begin
  cnt_process : process (clk_in, rst_l_in)
  begin
    if (rst_l_in = '0') then
      q      <= (others => '0');
      tc_out <= '0';
    elsif (clk_in'event and clk_in = '1') then
      if ld_in = '1' then
        q <= d_in;
      else
        if (en_tc_in = '1') and (en_in = '1') then
          q <= q + X"0001";
          if (q = x"FFFE") then
            tc_out <= '1';
          else
            tc_out <= '0';
          end if;
        end if;
      end if;
    end if;
  end process cnt_process;
  q_out <= q;
end rtl;
```

Figure 58: Sixteen-Bit Counter

Once the constraints are set, the synthesis and implementation are run. Running synthesis and implementation is usually a single click in the design tool. If the constraints are reasonable, the design should pass all checks. The inputs are not clocked in this design, so there may be timing violations on the inputs that require Flip-Flops to be added to the inputs. If there are no timing violations, tightening constraints like the clock frequency and rerunning the synthesis and implementation steps will show the FPGAs realistic performance. These tools produce several reports, like utilization, power, and timing. Use this opportunity to become familiar with these reports while they are small. Once the actual design is in place, the reports will contain a large amount of information and be difficult to parse through. In addition, it would be helpful to put many copies of the counter into the FPGA to increase utilization. A design like this will quickly use up the pins while the internal utilization of the FPGA will remain low.

6.2 Hardware Staging

In addition to practicing using the compilation tools, staging the hardware with a simple design can also be valuable. Usually, a reference board for an FPGA has an oscillator, a push button used for resetting the logic, and an LED that can indicate a logic level. This exercise is generally considered to be a hardware version of Hello World[3]. To start this exercise, the pin locations of the three components described above will need to be identified, and constraints will need to be set. If, for example, the clock oscillator is running at 100 MHz running the counter at this frequency and connecting the most significant bit of the counter to the LED will yield a flash rate of:

$$\frac{100\,MHz}{2^{16}} = 1526\,Hz$$

This rate is too fast to be able to see flashing. To mitigate this, two counters are cascaded. Figure 59 shows a cascaded arrangement of two counters. A listing of the code showing two cascaded counters is shown in Figure 60. The clock and reset inputs are connected to both instances of the counter. The enable signals are set to a logical one except for the terminal count from the first counter to the second counter. The load signal is set to a logical zero along with the d inputs. If the tenth bit of the second counter will have a flash rate of:

$$\frac{100\,MHz}{2^{26}} = 1.5\,Hz$$

This rate will be much easier to observe. Once the cascaded counter code is written and compiled, it is loaded from the design system to the FPGA development board. This process is the same no matter what the utilization of the part is. The LED should start flashing as soon as the design is loaded and will stop while the reset button is pushed.

---

[3] Hello World is a program introduced by Brian Kernighan in the book "The C Programming Language" (1978) as a program used to demonstrate a programming language and its tool-chain.

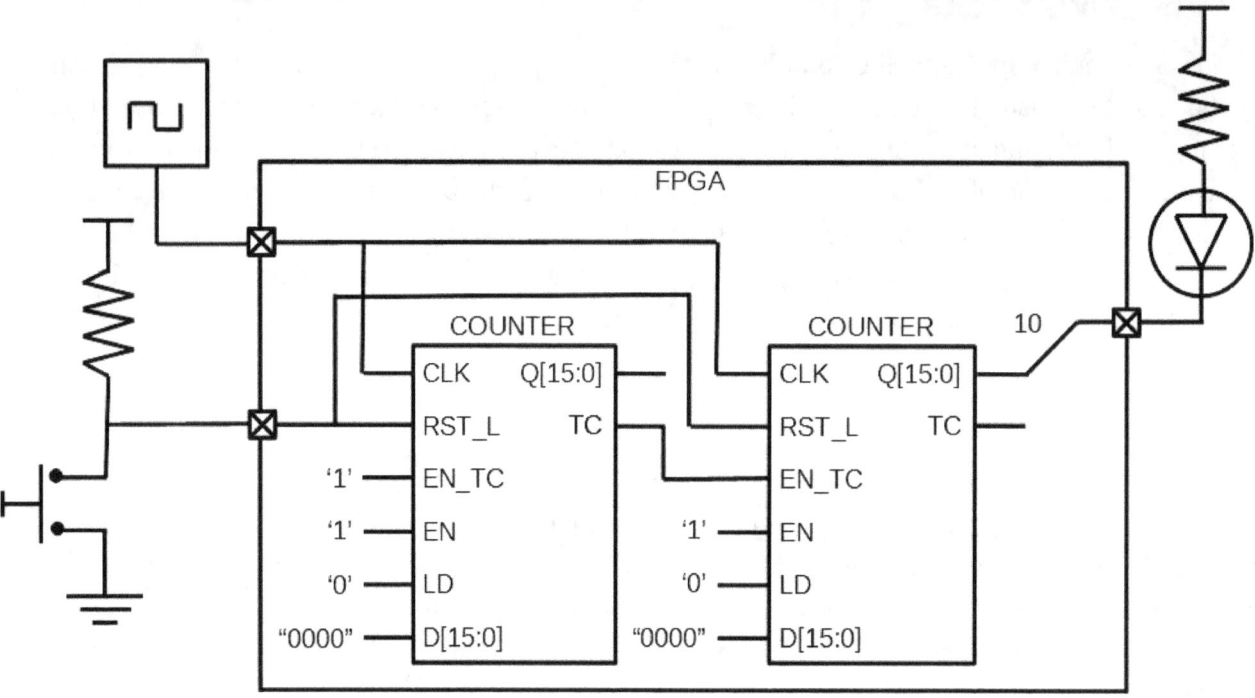

Figure 59: Cascaded Counters

The staging techniques shown in this section and the previous section are an excellent way to learn the operation of the tools and to ring out any anomalies before things are slowed down by a large design. The short time invested in this process will yield considerable timing savings during the development effort.

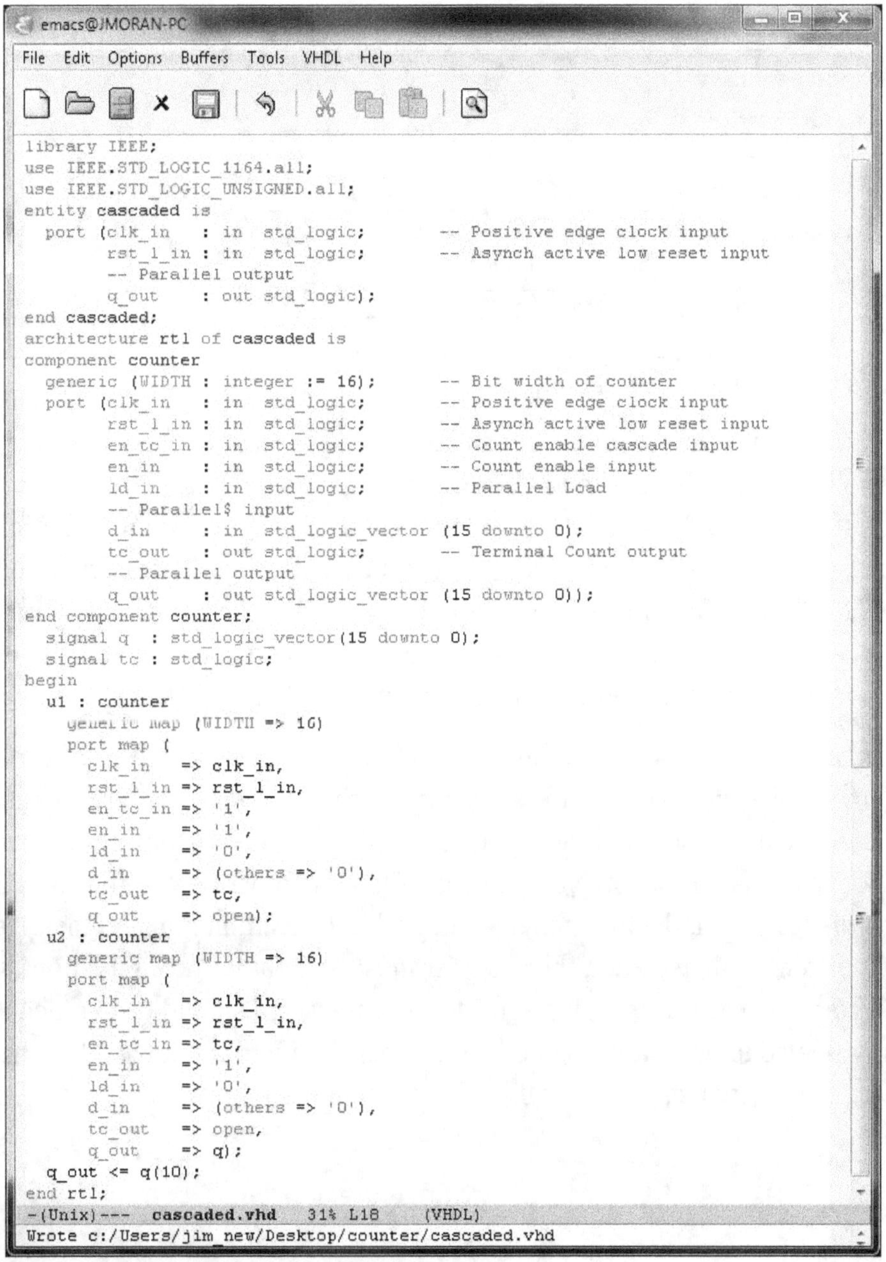

```vhdl
library IEEE;
use IEEE.STD_LOGIC_1164.all;
use IEEE.STD_LOGIC_UNSIGNED.all;
entity cascaded is
  port (clk_in   : in  std_logic;        -- Positive edge clock input
        rst_l_in : in  std_logic;        -- Asynch active low reset input
        -- Parallel output
        q_out    : out std_logic);
end cascaded;
architecture rtl of cascaded is
component counter
  generic (WIDTH : integer := 16);       -- Bit width of counter
  port (clk_in   : in  std_logic;        -- Positive edge clock input
        rst_l_in : in  std_logic;        -- Asynch active low reset input
        en_tc_in : in  std_logic;        -- Count enable cascade input
        en_in    : in  std_logic;        -- Count enable input
        ld_in    : in  std_logic;        -- Parallel Load
        -- Parallel$ input
        d_in     : in  std_logic_vector (15 downto 0);
        tc_out   : out std_logic;        -- Terminal Count output
        -- Parallel output
        q_out    : out std_logic_vector (15 downto 0));
end component counter;
  signal q  : std_logic_vector(15 downto 0);
  signal tc : std_logic;
begin
  u1 : counter
    generic map (WIDTH => 16)
    port map (
      clk_in   => clk_in,
      rst_l_in => rst_l_in,
      en_tc_in => '1',
      en_in    => '1',
      ld_in    => '0',
      d_in     => (others => '0'),
      tc_out   => tc,
      q_out    => open);
  u2 : counter
    generic map (WIDTH => 16)
    port map (
      clk_in   => clk_in,
      rst_l_in => rst_l_in,
      en_tc_in => tc,
      en_in    => '1',
      ld_in    => '0',
      d_in     => (others => '0'),
      tc_out   => open,
      q_out    => q);
  q_out <= q(10);
end rtl;
-(Unix)---  cascaded.vhd   31% L18    (VHDL)
Wrote c:/Users/jim_new/Desktop/counter/cascaded.vhd
```

Figure 60: Listing of Cascaded Counters

6.3 Simulation Staging

There are two levels of abstraction that are modeled in the simulation. Simulations at the RTL level abstraction simulate the code, as written, using a testbench. RTL simulations are generally used to verify the function of the code. Simulations at the gate-level abstraction are performed after place and route. Gate-level simulations verify the function of the implemented design and are used for dynamic timing analysis. A snippet of the simulation output of the sixteen-bit counter is shown in Figure 61.

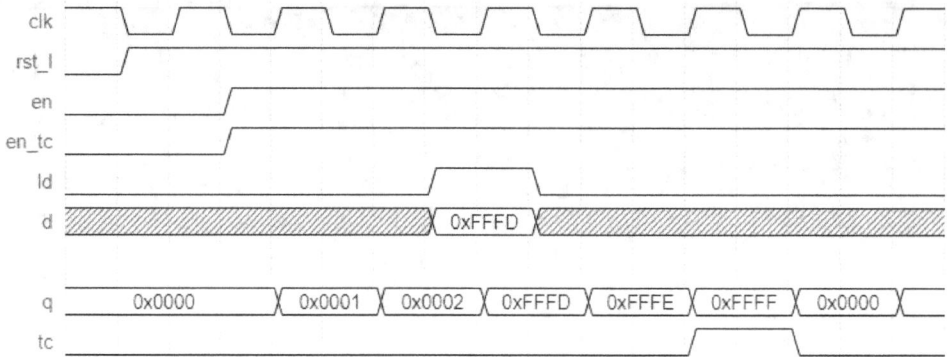
Figure 61: Counter Timing Diagram

Although verification is an independent topic, designers need to load the final FPGA design into a simulator and run a basic testbench. The FPGA design software initiates the process of loading the RTL design and testbench into the simulator. The testbench should, at a minimum, provide a stimulus for the clock, reset, and an initialization value for all inputs. A listing of a testbench for the counter is shown in Figure 62. Primary functions are tested at this point, but the value of performing this step during a staging operation is to ensure that all tools are in place and correctly linked. Observing reasonable results in the simulator is a good check that the simulation tools are working correctly.

The final step of simulation staging is gate-level simulation. In this case, the gate-level (post-implementation) model is simulated with the same testbench as the RTL model was run against. The gate-level simulation should yield the same result as the RTL simulation, except that the time delays should be more noticeable and probably varied. If good coding practice was followed in both the design model and the testbench, the functional result of the RTL simulation and the gate-level simulation should match. If they do not, the designer must find out why. Simulation-synthesis mismatches are a common cause of problems in FPGA designs and should be avoided at all costs. They are generally related to poor design practices like neglecting to initialize all of the Flip-Flops in a design, incomplete sensitivity lists, not handling clock domain crossings, or expecting memories to have initial values.

Figure 62: Counter Testbench Listing

6.4 Exercises

1) Redo the example shown in Chapter 6 using the development board identified in Exercise 2.1.

2) Insert an embedded logic analyzer and compare the results of Exercise 1 to the simulation results shown in Figure 61.

7. Case Studies: FPGA – ASIC Interactions

7.1 Using an FPGA for ASIC Verification

I have to admit I have not spent my whole career as an FPGA designer. About half my work is Application Specific Integrated Circuit (ASIC) design. With that said, I have done my share of FPGA prototyping during the ASIC design process. Some functions are difficult to verify using conventional simulation techniques because they take a prohibitive amount of simulation resources. These functions are frequently verified in an FPGA. FPGA prototyping can significantly accelerate the process of verifying ASIC functions. A fallacy associated with this technique is that following good design guidelines like those described in this text is unnecessary. After all, it is not production code, and it will never ship to a customer. Not treating this design as a true development can cause the verification effort to be unsuccessful and quite literally a waste of time. I fell into this trap during an ASIC development at a small startup.

This project was a unique ASIC development at the time. It was a bulk memory controller with a Reduced Instruction Set Computer (RISC) processor on the chip. By current standards, this would be called a System on a Chip (SoC) design. The verification effort was going well in verifying conventional logic functions on the ASIC. The problem was that it was nearly impossible to run any significant code on the RISC processor. Some small code loops were run that showed the processor was alive, but to verify the SoC, all of the processing features like communications, Direct Memory Access (DMA), and interrupts, as well as getting an idea of the performance, had to be verified. Targeting the ASIC code to an FPGA seemed to be an obvious choice.

The hardware platform for the verification effort was an off-the-shelf FPGA development board. Using an off-the-shelf platform circumvented the need to set most Spatial Constraints. The device manufacturer/family/device combination was selected, and the pins were fixed. The Temporal Constraints should be set now. The thought was that the ASIC technology was much faster than the FPGA technology, so the clock rate of the FPGA was set to one-fifth of the target clock rate of the ASIC. Reducing the clock rate seemed like an easy solution. In fact, so easy that there was no need to waste time on setting timing constraints. I learned a lesson here.

The ASIC design was also pretty big and would not fit in an FPGA, so some functions like error detection and correction for the memory and internal buffers were left out of the FPGA. The processor and associated peripheral code were loaded into an FPGA tool along with the manufacturer's Spatial Constraints for the development board. The FPGA was synthesized and implemented. There were a bunch of warnings, but with a few cleanups, there was an FPGA image ready to program into the FPGA. The executable code for the RISC processor was loaded into a memory on the board that the processor could access. The reset switch was pressed, and something happened, but not what I expected.

My reaction in this situation is to press the reset again. Something happened, not what was expected, and different than the first time. Try again? Albert Einstein defined insanity as "doing the same thing over and over again and expecting different results." I saw different results, so I'm not insane if I keep trying, right? Not exactly, I saw the same results, not working every time. The actual incorrect operation was not pertinent. After a few too many reset presses, I decided to be an engineer and analyze the problem. If you read Section 1.6, you should think this is a reset synchronization problem. You are half right. The actual reset signal on the switch was bouncing a lot. The reset synchronizer worked well but debouncing logic was needed to clean up the reset signal thoroughly. An example of debounce logic is shown in Figure 63.

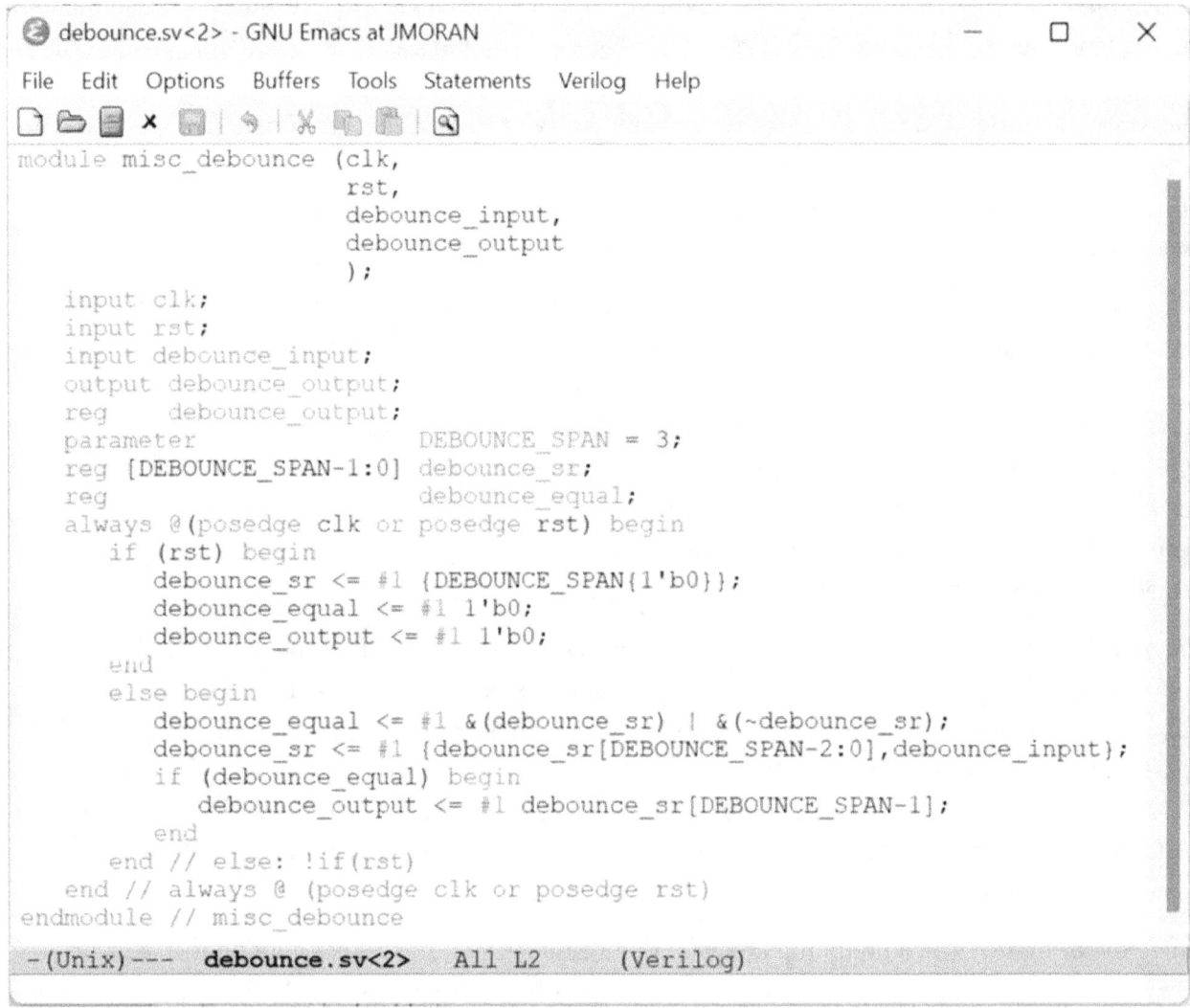

```verilog
module misc_debounce (clk,
                      rst,
                      debounce_input,
                      debounce_output
                      );
   input clk;
   input rst;
   input debounce_input;
   output debounce_output;
   reg    debounce_output;
   parameter             DEBOUNCE_SPAN = 3;
   reg [DEBOUNCE_SPAN-1:0] debounce_sr;
   reg                   debounce_equal;
   always @(posedge clk or posedge rst) begin
      if (rst) begin
         debounce_sr <= #1 {DEBOUNCE_SPAN{1'b0}};
         debounce_equal <= #1 1'b0;
         debounce_output <= #1 1'b0;
      end
      else begin
         debounce_equal <= #1 &(debounce_sr) | &(~debounce_sr);
         debounce_sr <= #1 {debounce_sr[DEBOUNCE_SPAN-2:0],debounce_input};
         if (debounce_equal) begin
            debounce_output <= #1 debounce_sr[DEBOUNCE_SPAN-1];
         end
      end // else: !if(rst)
   end // always @ (posedge clk or posedge rst)
endmodule // misc_debounce
```

Figure 63: Verilog Code for Debouncing

Now the system starts up the same way every time, still wrong though. More investigation showed that the memory contents are not correct. The code got written into the memory incorrectly after programming, but a read of the contents shows consistently wrong results. At least it's consistently wrong now. This memory error points to a memory write problem. Without getting too deep into the process, this turned out to be a small clock skew problem and a significant timing closure problem. Because there were no timing constraints, the implementation tools took the easiest path to implement the design. Setting proper timing constraints did not fix the problem. It was never anticipated that this code would be connected to the pins of an FPGA. There was logic after the last Flip-Flop, and the place and route tool could not put the Flip-Flip in the IO block (See Figure 2 on Page 14). Fixing that allowed the tool to produce an implementation that would achieve timing closure. Fixing a few additional annoyances allowed the design to work as initially anticipated.

You might be wondering if this was worth it. The answer is "kind of". A lot of time went by to fix the problems. This was time that could have been spent developing actual mission code. Because of the elapsed time, more verification was done, and any hardware bugs had been found. A platform to run the RISC code on gave the software engineers a test bed to debug the mission software. Although this was not the primary intent, it positively contributed to getting the whole system verified and fielded. There was an unintended advantage, though.

A verification engineer helping me through the process was also on the hook to verify the error detection and correction logic. Error detection and correction algorithms, commonly used in memory systems, are very prime, meaning the codes cannot be factored. Any codes that are factorable can cause memory errors to escape. Because of this, verifying them without using a brute force method of trying all combinations of errors with every possible data pattern is difficult. Trying every possible data pattern is prohibitive in simulation. Remember that I left this out of the RISC processor test platform. The verification engineer saw this as an opportunity. He could put the error detection and correction algorithm on an FPGA with a data generator and checker. Run it at the fastest clock frequency the FPGA could support, and try all combinations.

A quick calculation showed that this would still take many days. It did not fill the FPGA, though. He put about ten copies covering ten different spaces of the algorithm and set it to run. He left it for the weekend. This was a proprietary algorithm. More efficient than the state of the art and a selling point for the system. We came in on Monday morning, and it had completed. We expected it to show that it passed, but it had found a problem. A bug in the algorithm allowed a specific type of error to escape. With some work to isolate what failed, we gave the data to the algorithm designer. He figured out the flaw and changed the algorithm accordingly. By now, it was Friday again. The test system was started, and on Monday morning, the test was completed and passed. This bug would have been

a flaw in the algorithm we marketed as a unique feature. It would not have been found in simulation. The FPGA validation quickly paid for itself by finding this single bug.

7.2 FPGA Design for ASIC Retargeting

I have also done my share of FPGA designs that were intended to be retargeted to an ASIC. FPGAs are expensive, too expensive to ship on products sensitive to profit margin. ASICs are significantly less expensive per piece. ASICs have significant Non-recurring Engineering (NRE) costs, but their piece price is low. Additionally, ASICs are not reconfigurable. If, after they are built, a bug is found or the design specification changes, they have to be respun. The NRE for the respin is nearly the same as the first run and the respin takes as long as the first pass. FPGAs make excellent platforms for prototyping and early production. Most bugs and design specification changes are identified at this point in the process. If FPGAs are used for early production, the product is brought to market earlier, potentially beating the competition and gaining market share. Whatever the motivation for retargeting, the FPGA design must be done with ASIC retargeting in mind. There is a lot of documentation from ASIC manufacturers, FPGA manufacturers, and independent parties that describe steps to be taken when designing an FPGA that will eventually be retargeted to an ASIC. A lot of the information in these documents is specific to the respective manufacturer's components or the application being designed. Here are a few general points:

- Never let the delay of the FPGA enable its operation. For example, if you rely on the delay of a path in the FPGA to cause the signal change on a later clock cycle, it is not guaranteed that the ASIC will have that much delay. The ASIC is almost always much faster than the FPGA. Relying on the delay of paths in a digital design is not good design practice.

- Do not use a feature on the FPGA that is not available in the ASIC. The most obvious feature is reconfigurability. If the reconfigurable nature of the FPGA is used in the system's mission, it cannot be retargeted as an ASIC. There are some very limited cases where you can reconfigure small parts of an ASIC, but you need to identify this early in the process. Other features in FPGAs can be retargeted to an ASIC, but they require custom work in the retargeting process. These functions include memories, clock managers, and digital signal processing accelerators. Also, System on Chip (SoC) FPGAs are complicated to retarget both because of the complexity of the special functions and the potential cost of intellectual property that would need to be purchased for the ASIC design.

- The IO technology and packaging options may be much different on the FPGA than those available for an ASIC. It is essential to research and compare options early in the process. Comparing options early in the process will make the retargeting process straightforward and minimize the impact on the PCB design.

An example showing a cost comparison between an FPGAs and ASICs based on quantity is outlined in Table 5. It indicates that the NRE for an FPGA is a mere $10,000. This covers the cost of two copies of the design software and two computers to run the software on. The ASIC NRE is 100 times more at approximately $1,000,000. Most of the NRE cost is in the masks that must be produced for the fabrication process. These masks have to be remade if the ASIC must be respun. The piece price is the opposite. A single large FPGA can cost $5000, which is 100 times more than the projected ASIC piece price of $50.

Table 5: FPGA - ASIC Cost Comparison

	FPGA	ASIC
NRE	$10,000	$1,000,000
Piece Price	$5,000	$50
Quantity	**Total Cost**	
50	$260,000	$1,002,500
100	$510,000	$1,005,000
200	$1,010,000	$1,010,000
500	$2,510,000	$1,025,000

The break-even point between FPGAs and ASICs is illustrated in Figure 64. The chart shows that for this example, the break-even point is 200 pieces. This quantity would be a pretty small production run, and it is evident that FPGAs are only practical for small production quantities. I should also mention that the lead time for an ASIC could easily be three to six months. Other advantages to having FPGAs available for initial production are that the design is validated at the system level, some experience is gained from actual system use, and early adopting customers can access your product.

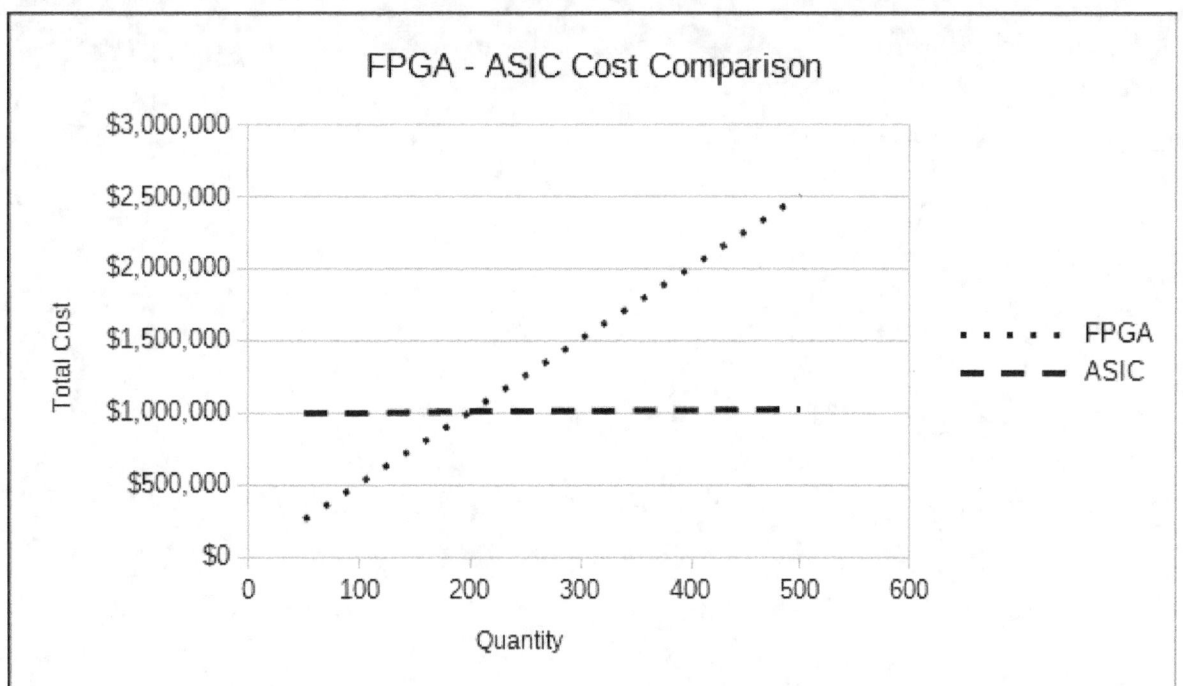

Figure 64: FPGA - ASIC Cost Comparison

7.3 Conditional Compilation

In cases where the design is likely to be retargeted from an FPGA to an ASIC or vice versa, it is best to make as few changes as possible. There are constructs in both VHDL and SystemVerilog that allow for conditional compilation. This concept enables multiple models of a function to be included in a design file, but only one model is compiled or simulated at any given time. The example in Figure 65 is a model of a static RAM contained in a wrapper. If the RAM_TYPE is set to FPGA, the FPGA model is compiled and simulated. If the RAM_TYPE is set to ASIC, the ASIC model is compiled and simulated. Finally, if neither is selected, a generic HDL model is used. This HDL may be synthesizable, but most likely, the generic model will be used for simulation only.

All models must be functionally equivalent. In this example, the write enable signal is different between models. It is active high in the ASIC model and active low in the FPGA model. To mitigate this difference, an inverter is inserted in the FPGA model. The generic model is designed to be active high to match the input. All models are used in the design that pertain to the ASIC should use the ASIC value for the type and all models that pertain to the FPGA should use the FPGA value for the type. This assures that the whole design compiles for the given target without any changes to the code except for the type value. This is true for simulation as well.

It is best to plan for this eventuality before starting the design because it can be difficult to retrofit the code. It must be verified that the type value is set the same for compilation and simulation.

Figure 65: Conditional Compile Example

7.4 Exercise

1) Use the example described in Figure 53 on page 67 to get quotes for both the FPGA component and an equivalent ASIC component. Use the quantities listed in Table 5 and find the break-even point.

Conclusion

A well-written and verified code base based on a single source of truth is essential. We have seen that it is only the foundation of a successful design. Properly constraining the design is a crucial step in developing a reliable and producible design. Simply constraining the design does not necessarily complete the design either. It only defines "the box". Making time versus space trade-offs to stay within these constraints yields a final product that meets your performance, budget, and time-to-market goals.

This book lays out steps for successful FPGA design once the logic design is complete. The book is not complete. There are always special cases that are unique to the specific application. Using the experience gained in this book and in practice, you should be able to identify these special cases and deal with them appropriately.

Also, the reader has probably noticed a lack of manufacturer-specific examples in this book. This is on purpose. Manufacturers frequently change their components and design software rendering any examples contained in a static text quickly obsolete. The concepts of Temporal Constraints and Spatial Constraints remain the same, though. Understanding these constraints allows you to effectively use any components and design software a manufacturer offers. Take the manufacturer's training for their development software to learn the specifics of the tools, like the language for setting constraints, not to understand why constraints are essential or how to use them.

I want to tell you that reading and understanding this book will save you some sleep during the process of bringing the FPGA-based printed circuit board up in the lab. The initial bring-up is a team effort involving the FPGA designers, printed circuit board designers, software engineers, and other subject matter experts. There are always problems. If you read, understand, and apply the concepts in this book, you will be satisfied with helping other disciplines find and fix their problems instead of needing to fix your own problems.

Solutions

1.1) $F_{MAX} = \dfrac{1}{(150\,pS + 170\,pS + 60\,pS)} = 2.63\,GHz$

1.2) Slack = 400 pS – 380 pS = 20 pS, timing closure is achieved.

1.3) $F_{MAX} = \dfrac{1}{(150\,pS + 170\,pS + 60\,pS - 90\,pS)} = 3.45\,GHz$

1.4) Slack = 400 pS – 290 pS = 110 pS, timing closure is achieved.

1.5) $F_{MAX} = \dfrac{1}{(150\,pS + 170\,pS + 60\,pS + 90\,ps)} = 2.13\,GHz$

1.6) Slack = 400 pS – 470 pS = -70 pS, timing closure is not achieved.

1.7) Skew$_{MAX}$ = 400 pS – 230 pS – 0 pS = 170 pS

1.8) F_{MAX} = 2.13 GHz (The slowest frequency defines the F_{MAX}.

1.9) $F_{MAX} = \dfrac{1}{(150\,pS + 60\,pS + 90\,pS)} = 3.33\,GHz$

1.10) Figure 66 is a timing diagram with the skews shown in the exercise. It can be seen that the slack between U2 and U3 is smaller than the slack between U1 and U2. Because of this, the path from U2 to U3 is going to define F_{MAX}.

$$F_{MAX} = \frac{1}{(Tck\,2\,q + Tp\,2 + Tsu + (Tp\,5 - Tp\,4))} = \frac{1}{(250\,pS + 150\,pS + 200\,pS + (300\,pS - 200\,pS))} = 1.25\,GHz$$

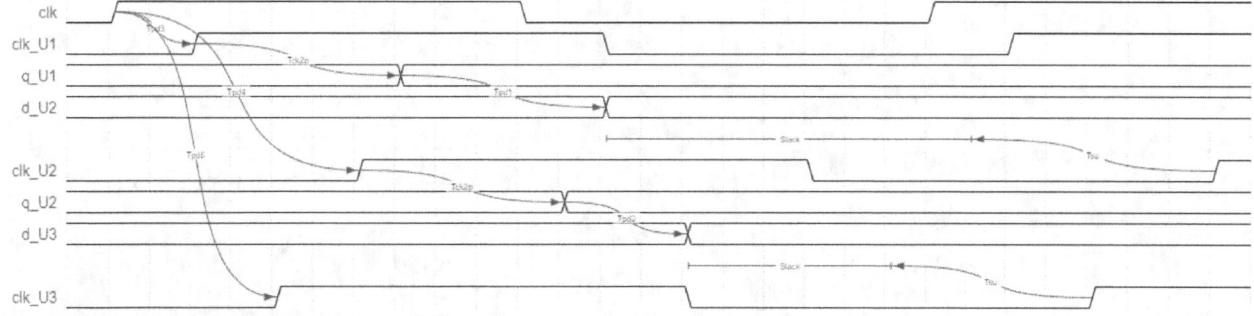

Figure 66: Multiple Skew Solution

1.11) Figure 67shows the additional flip-flops for a safe clock domain crossing, U1a and U2a need to be added because logic isn't allowed in the clock domain crossing.

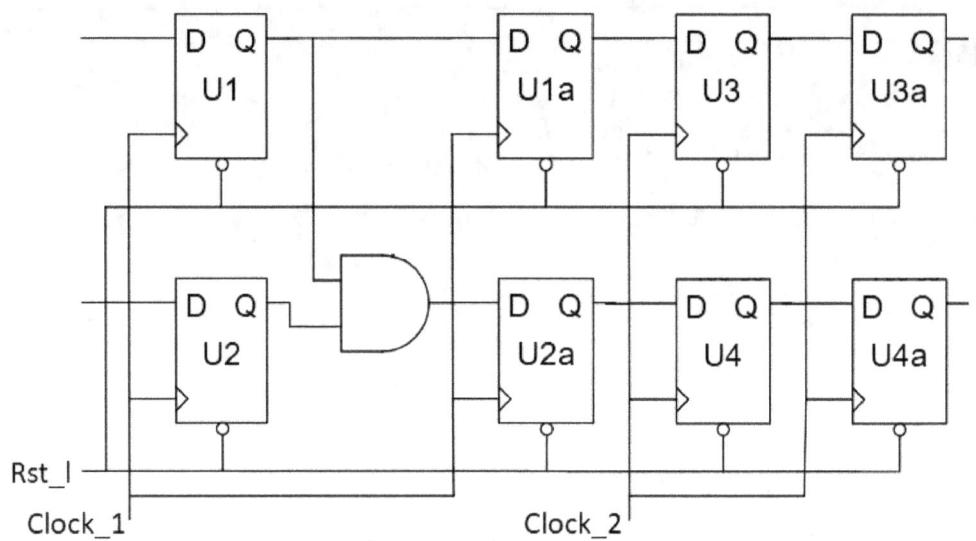

Figure 67: Synchronizer with Logic

1.12) Figure 68 is an example solution. Other solutions are possible.

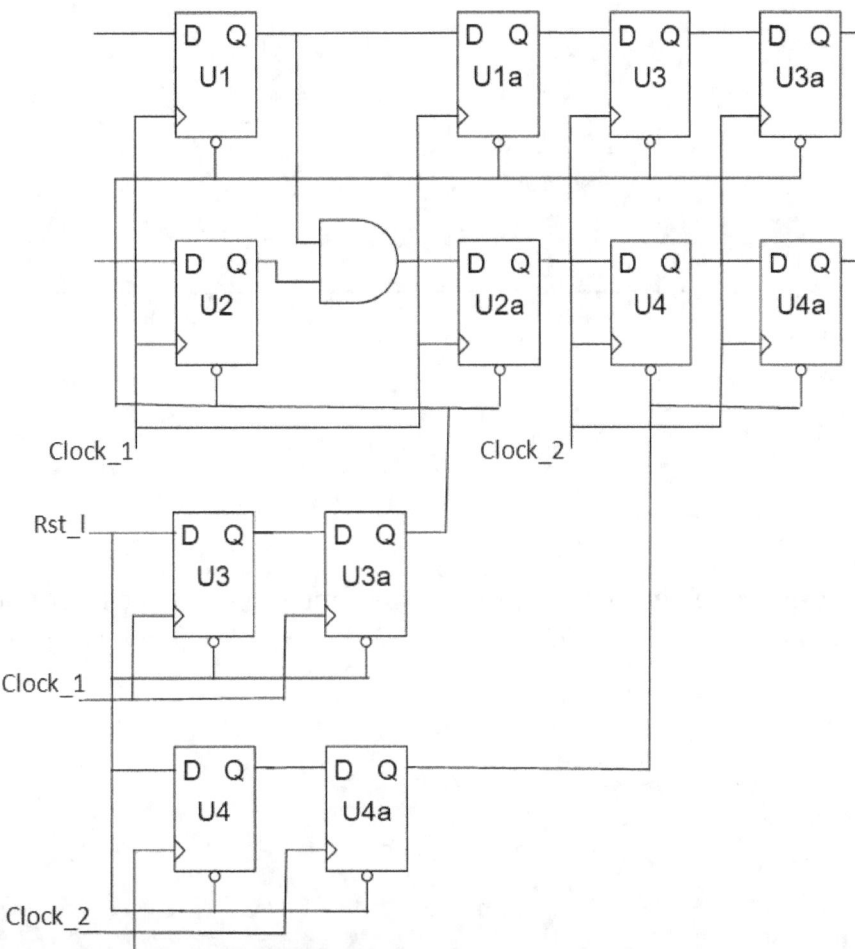

Figure 68: Exercise 1.12 Solution

1.13) No, there is still a delay from the clock input to the divided clock output.

1.14)$F_{MAX} = \dfrac{1}{(150\,pS + 170\,pS + 80\,pS + 100\,pS + 60\,pS)} = 1.79\,GHz$

Latency = 2*(150pS + 170pS + 80pS +100 pS + 60pS) = 1.12 nS

1.15) $F_{MAX} = \dfrac{1}{(150\,pS + 170\,pS + 60\,pS)} = 2.63\,GHz$

Latency = 4*(150pS + 170pS + 60pS) = 1.52 nS

1.16) Figure 69 is the smallest arrangement that is faster than 2.5GHz

$F_{MAX} = \dfrac{1}{(150\,pS + (100\,pS + 80\,pS) + 60\,pS)} = 2.56\,GHz$

Latency = 3*(150pS + 180pS + 60pS) = 1.17 nS

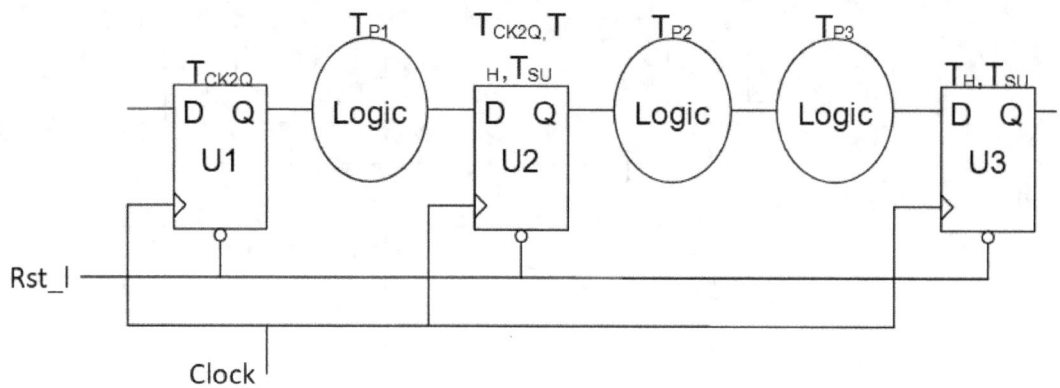

Figure 69: Exercise 1.17 Solution

2.1) Various solutions.

2.2) Using routing similar to Figure 70 all signals can be routed on a single layer. Because there needs to be an even number of signals layers plus one power plane and one ground plane the total layer count is 4.

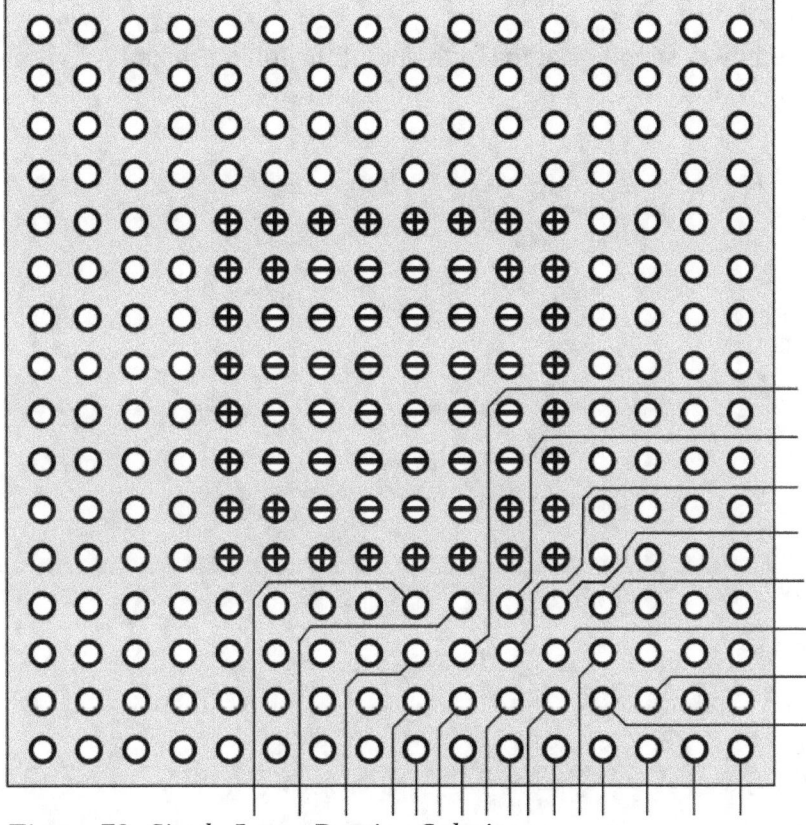

Figure 70: Single Layer Routing Solution

2.3) The routing solution is similar to Figure 34 on Page 41 where all routes in a column need to go through the same set of balls. This requires 3 layers. Because there needs to be an even number of signals layers plus one power plane and one ground plane the total layer count is 6.

2.4) U1-U2

Low Noise Margin $\quad = V_{IL} - V_{OL} = 0.7V - 0.5V = 0.2V;$

High Noise Margin $\quad = V_{OH} - V_{IH}$
$\qquad = 1.8V - (3.3V*.95 - 1.5V)$ (Use the low range of the power supply tolerance)
$\qquad = 1.8V - (3.135V - 1.5V)$
$\qquad = 0.165V$

U2->U1

Low Noise Margin $\quad = V_{IL} - V_{OL} = 0.7V - 0.5V = 0.2V;$

High Noise Margin $\quad = V_{OH} - V_{IH} =$
$\qquad = 2.6V - (2.5V*.95 - 0.5V)$ (Use the low range of the power supply tolerance)
$\qquad = 2.6V - (2.375V - 0.5V)$
$\qquad = 0.725V$

Max limit $= V_{IH}$ Max $- V_{OH}$ Max $\qquad = V_{CCIO}*0.95 + 1V - V_{CCIO}*1.05$
$\qquad\qquad = 3.375V - 3.465V$

$$= -0.09V$$

A violation if the 3.3V power supply is at the high end of the tolerance and 2.5V power supply is at the low end.

3.1) Various Solutions

4.1) Various Solutions

4.2) Various Solutions

4.3) Various Solutions

4.4)Various Solutions

5.1) Various Solutions

5.2) Various Solutions

5.3) Various Solutions

5.4) Various Solutions

6.1) Various Solutions

6.2) Various Solutions

7.1) Various Solutions

Bibliography

Bergeron, Janick. *Writing Testbenches Using SystemVerilog*, New York: Springer, 2006.

Jasinski, Ricardo. Effective coding with VHDL: principles and practice. Cambridge, MA: MIT Press, 2016.

Pedroni, Volnei A. *Circuit Design with VHDL*, 3rd ed. Cambridge, MA: MIT Press, 2020.

Somerville, Ian. *Software Engineering*, 10th ed. Essex, England: Pearson, 2015.

Sutherland, Stuart. *RTL Modeling with SystemVerilog for Simulation and Synthesis: Using SystemVerilog for ASIC and FPGA Design* 1st ed. Tualatin, OR: Sutherland HDL, Inc., 2017.

Index